Hands on Learning in Machine Learning and Deep Learning

The Author

Dr. Vinod Kumar Yadav is Senior Scientist (Agricultural Statistics) at Fisheries Economics, Extension and Statistics (FEES) Division, ICAR-Central Institute of Fisheries Education (CIFE) Mumbai. He has more than 15 years of teaching and research Experience. His research interest includes Modelling and Forecasting, GIS and Remote Sensing, Climate change, Artificial Intelligence and Big Data Analytics. He is teaching several courses at Master and Ph.D. level like Mathematics, Statistics Methods. Advanced statistical Methods, GIS and Remote Sensing, Artificial Intelligence and Big data Analytics. Dr. Yadav received Master degree from Indian Agricultural Research Institute (IARI) Pusa New Delhi and Ph.D. from IIT Bombay.

Hands on Learning in Machine Learning and Deep Learning

– Author –

Dr. Vinod Kumar Yadav

PhD (IIT Bombay)
Senior Scientist,
Fisheries Economics, Extension and Statistics Division,
ICAR–Central Institute of Fisheries Education (CIFE), Mumbai

2026

Associated Publishing Company
A Division of
Astral International Pvt. Ltd.
New Delhi – 110 002

ISBN : 978-93-5919-691-6

Published by : **Associated Publishing Company**
A Division of
Astral International Pvt. Ltd.
– ISO 9001: 2015 Certified Company –
4736/23, Ground Floor, Ansari Road
Darya Ganj, New Delhi-110 002
Ph. 011-43549197, 23278134
e-mail: info@astralint.com
Website: www.astralint.com

Printed at : **Replika Press Pvt Ltd**

Preface

Knowledge in the field of Computer Vision and Machine Learning is advancing at a rapid pace, leading to new and varied applications. The ace of Computer Vision and Machine Learning tools has now touched every field of science and management. Several applications of these technologies are possible in the fisheries and aquaculture sector, too. So, the fisheries development officers, entrepreneurs, farmers and fisheries professionals also need to be equipped with the advanced edge of the technological revolution.

Realizing the importance of this subject, a book on " Hands-on learning in Machine Learning and Deep Learning" was conceptualized for research scholars, fisheries development officers, entrepreneurs, technical experts and fish farmers who have an interest in Machine Learning technology. The pedagogy of this book involved theory and practical sessions to provide comprehensive learning and thus provide complete theoretical and practical knowledge about different Machine learning algorithms, object detection, object classification, and feature extraction. Additionally, Practical experience with YoLo, R-software, PyCharm software and Google Colab environment was also provided.

Under this book, different open-source software was selected for practical exercises.

Vinod Kumar Yadav

Contents

Chapter 1

Introduction to Python and it's Installation

What is Python?

Python is a popular programming language. It was created by Guido van Rossum, and released in 1991.

It is used for:

- web development (server-side),
- software development,
- mathematics,
- system scripting.

What can Python do?

- Python can be used on a server to create web applications.
- Python can be used alongside software to create workflows.
- Python can connect to database systems. It can also read and modify files.
- Python can be used to handle big data and perform complex mathematics.
- Python can be used for rapid prototyping, or for production-ready software development.

Why Python?

- Python works on different platforms (Windows, Mac, Linux, Raspberry Pi, *etc.*).

- Python has a simple syntax similar to the English language.
- Python has a syntax that allows developers to write programs with fewer lines than some other programming languages.
- Python runs on an interpreter system, meaning that code can be executed as soon as it is written. This means that prototyping can be rapid.
- Python can be treated procedurally, in an object-oriented way or in a functional way.

Python Environment

Here, for the practical sessions, we will use Python version 3.8.10. We will be using two Python platforms – PyCharm IDE and Google Collab. The environments chosen are quite simple and easy to adapt. We will first install the PyCharm and then acquaint ourselves with the Google Colab virtual environment.

Installation of Python Environment in Windows 11

For running different Python platforms like Pycharm, Visual Studio, IDLE, Anaconda, *etc.* It is necessary to have Python installed. Henceforth, the proper step-by-step guide for the installation of Python.

Step 1: Download Python 3.8.10

- Open your web browser and go to the official Python website: https: // www.python.org/.
- This option allows you to see all available versions of Python.
- Scroll down to the section "Looking for a specific release?"
- This directs you to the specific release page for Python 3.8.10.

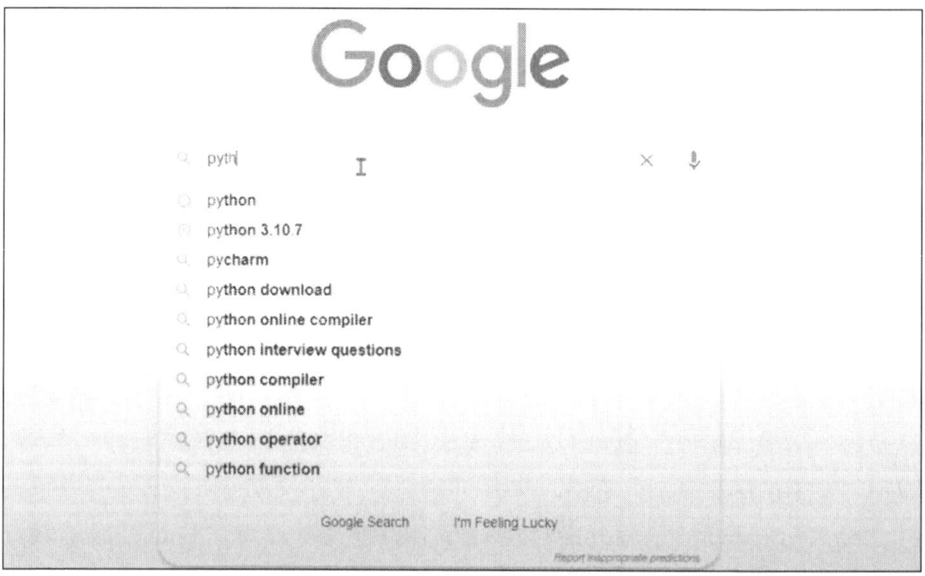

☆ This is the official site where Python releases are hosted.

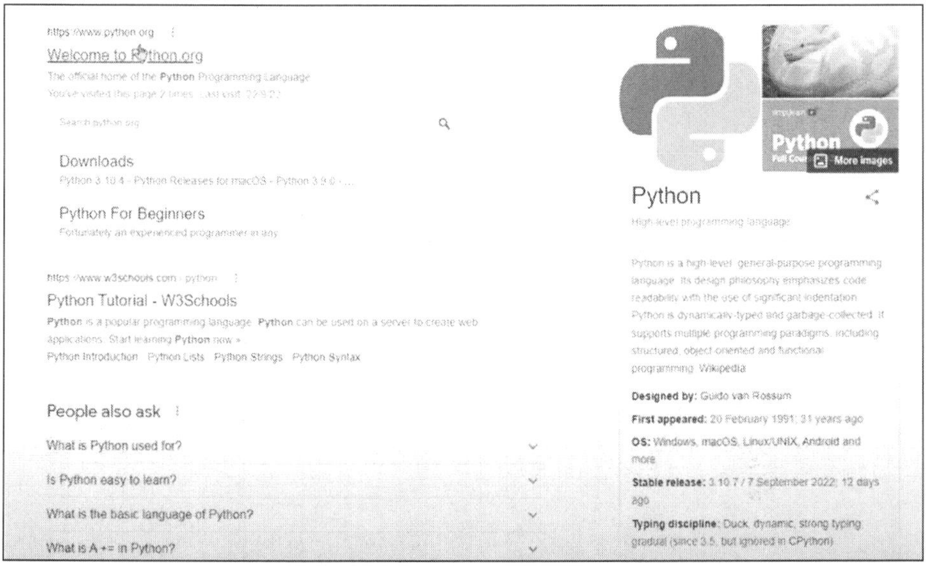

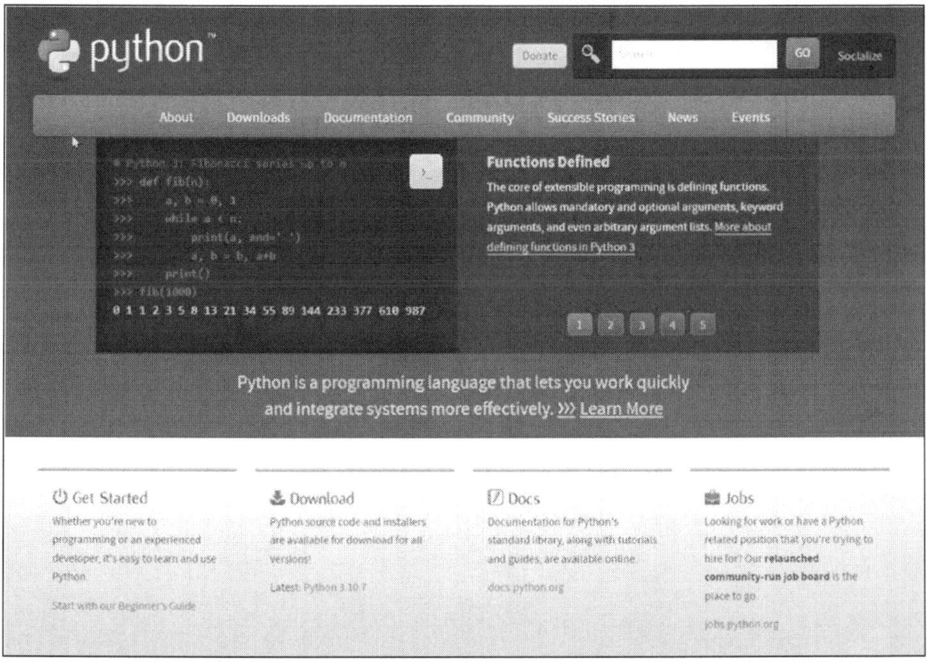

- ☆ Hover over the "Downloads" tab and click on "View the full list of downloads" at the bottom of the dropdown.

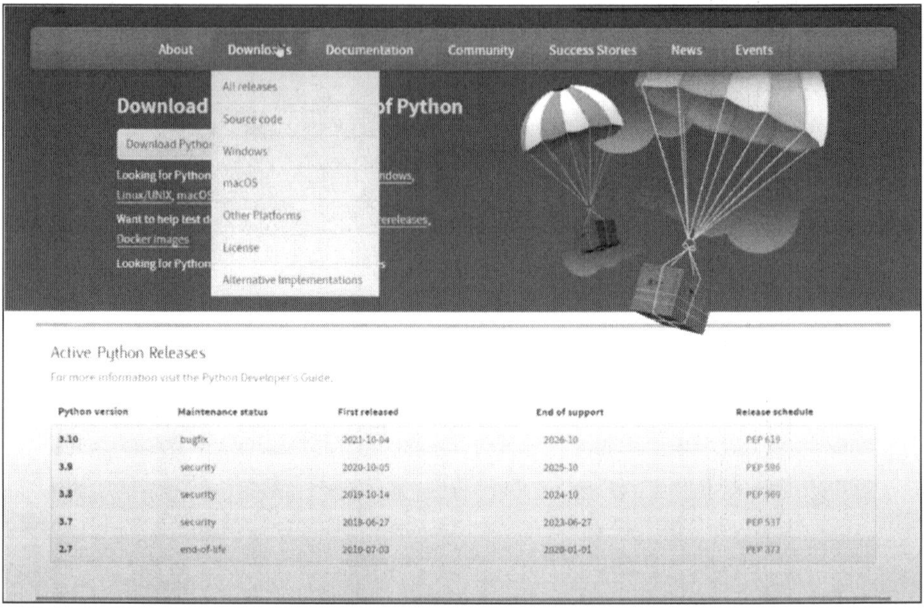

- ☆ This section contains links to all Python releases, including old versions.
- ☆ Click on "Python 3.8.10" under the "Python 3.8.10 - May 3, 2021" section.
- ☆ Scroll down to the "Files" section and click on "Windows x86-64 executable installer" to download the installer for 64-bit Windows.
- ☆ This file is the executable installer for the 64-bit version of Python, suitable for most modern Windows systems.

Locate the Installer

☆ Once the download is complete, locate the downloaded file (usually in the "Downloads" folder).

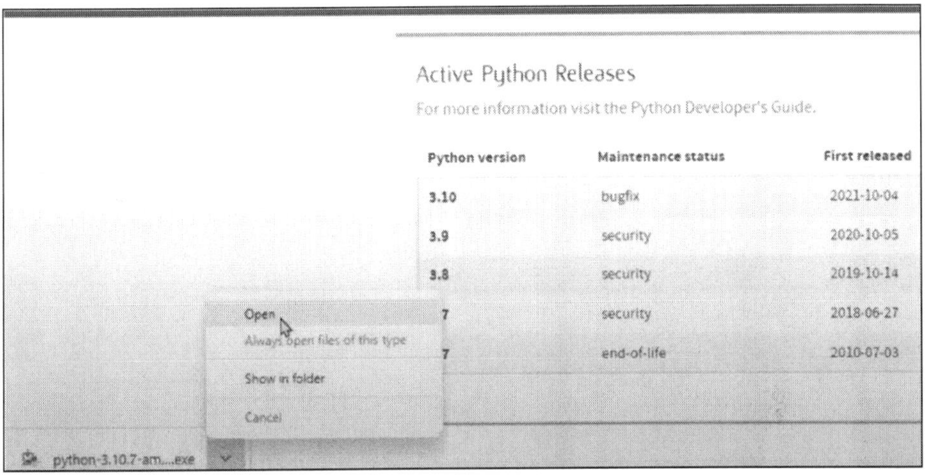

Run the Installer

☆ Double-click on the downloaded file to run the installer.

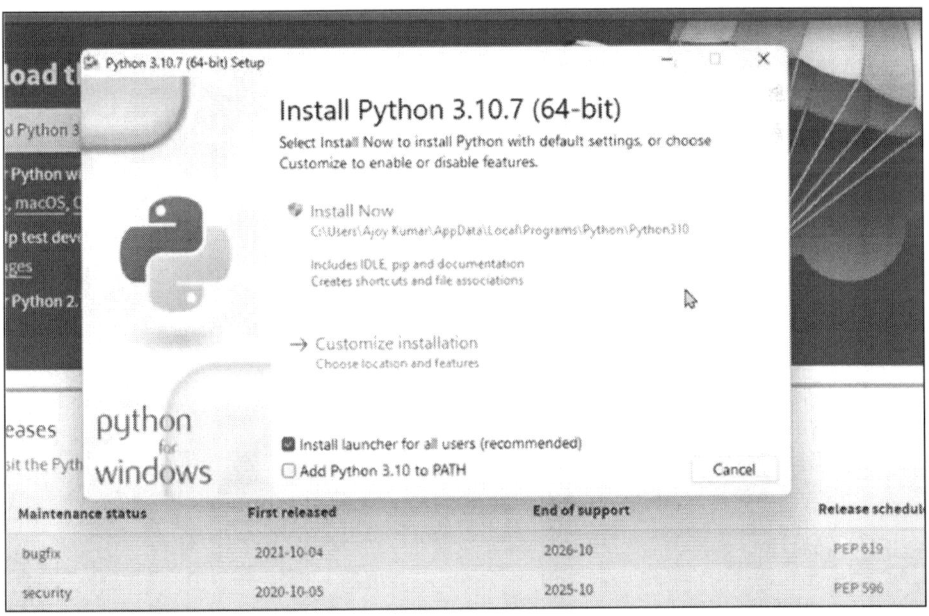

Setup Options

- **Important:** Check the box that says "Add Python 3.8 to PATH." This will make it easier to run Python from the command line.
- Click on "Customize installation."

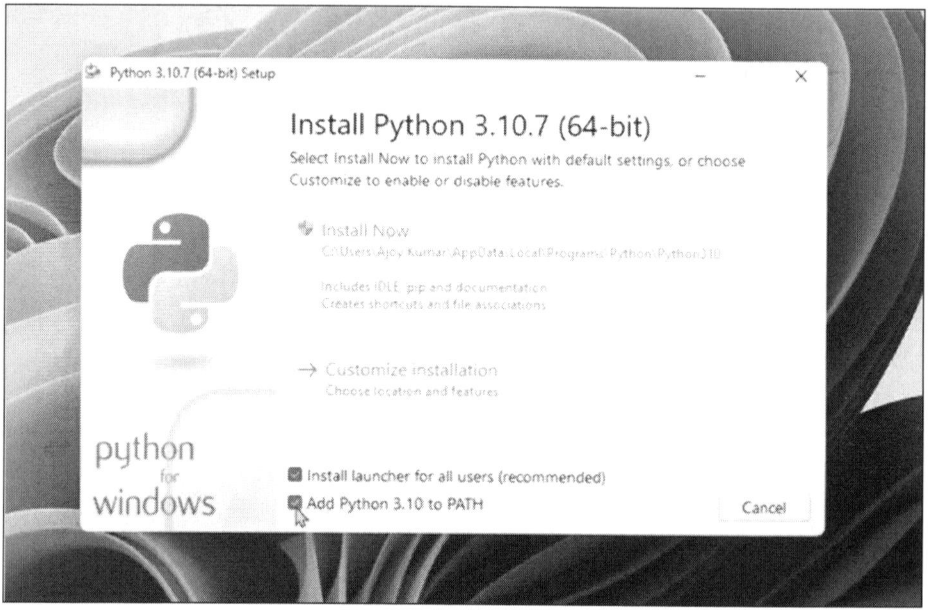

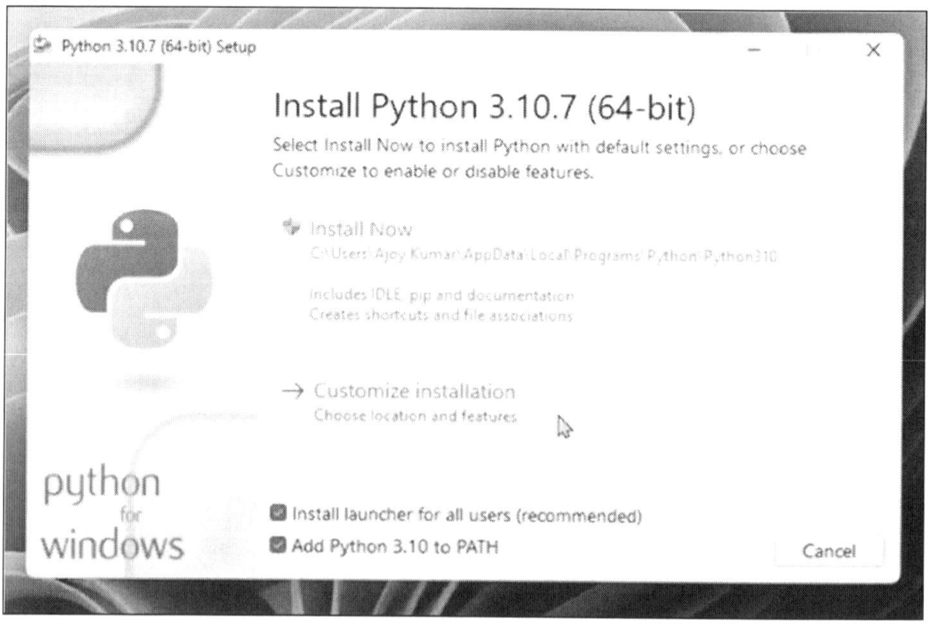

Optional Features

☆ Ensure that all options are checked (this includes Documentation, pip, tcl/tk and IDLE, and others). Click "Next."

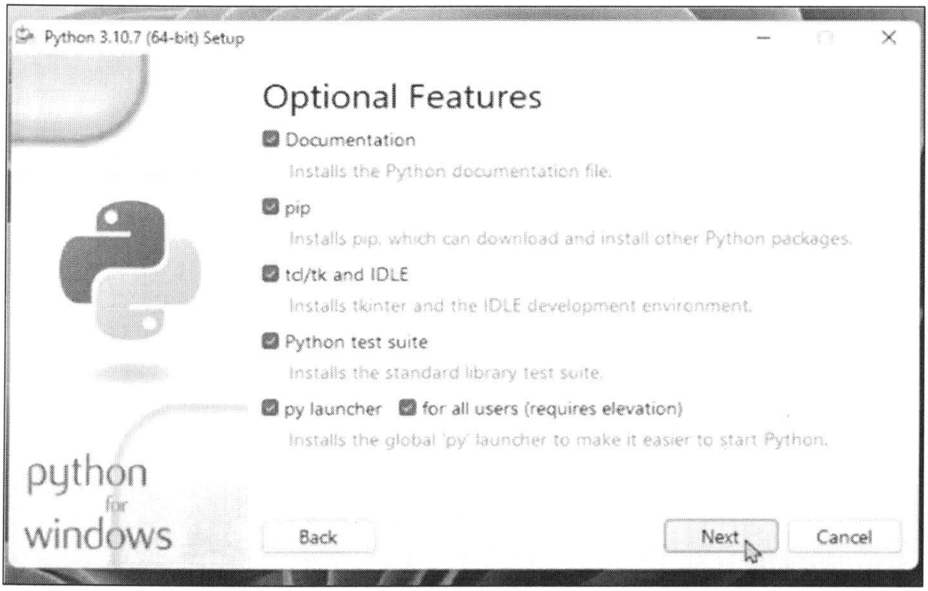

Advanced Options

☆ Check the boxes for "Install for all users" and "Precompile standard library."

☆ You can leave the install location as default or change it if necessary.

☆ Click "Install."

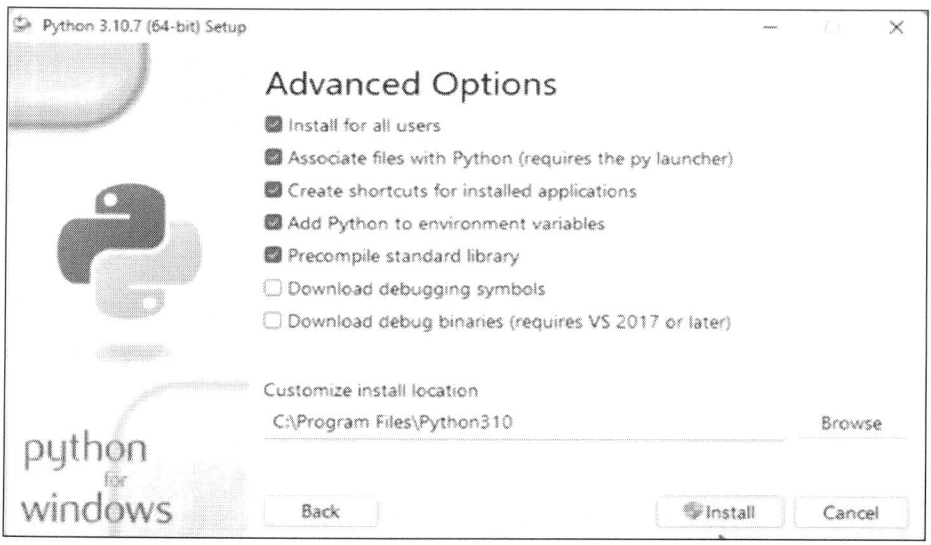

Installation Progress

☆ Wait for the installation process to complete. This may take a few minutes.

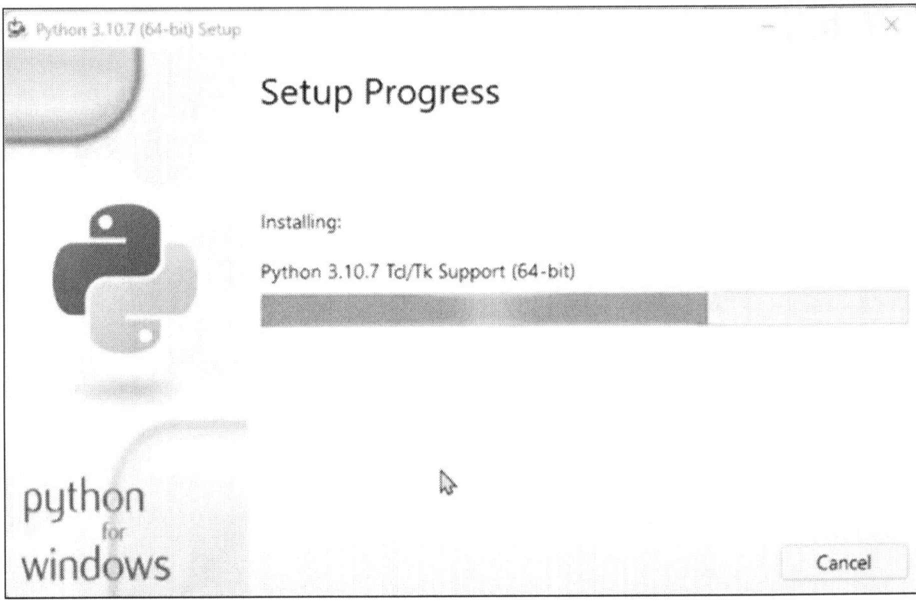

Completion

☆ Once the installation is complete, you should see a "Setup was successful" message. Click "Close" to exit the installer.

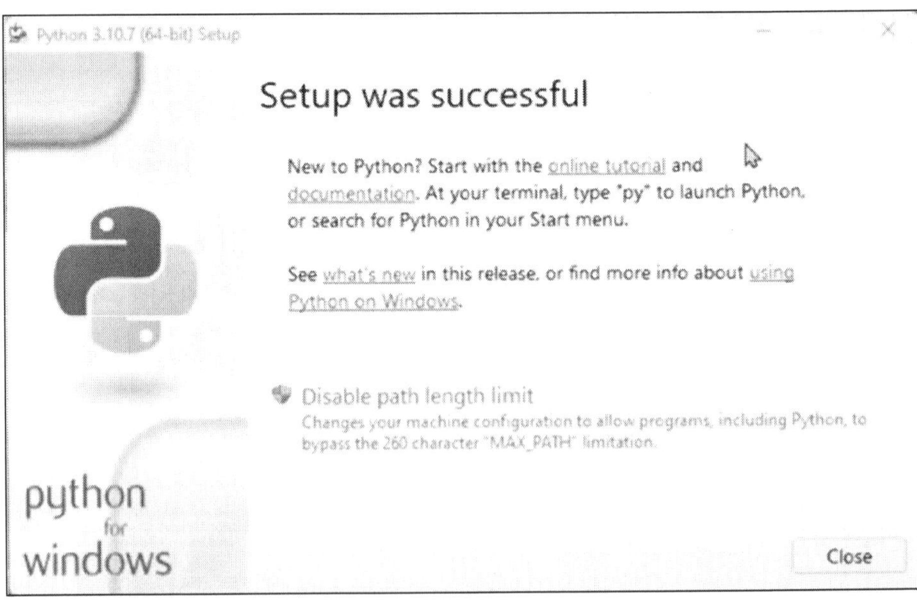

Open Command Prompt

- ☆ Press Win + R, type cmd, and press Enter to open the Command Prompt.

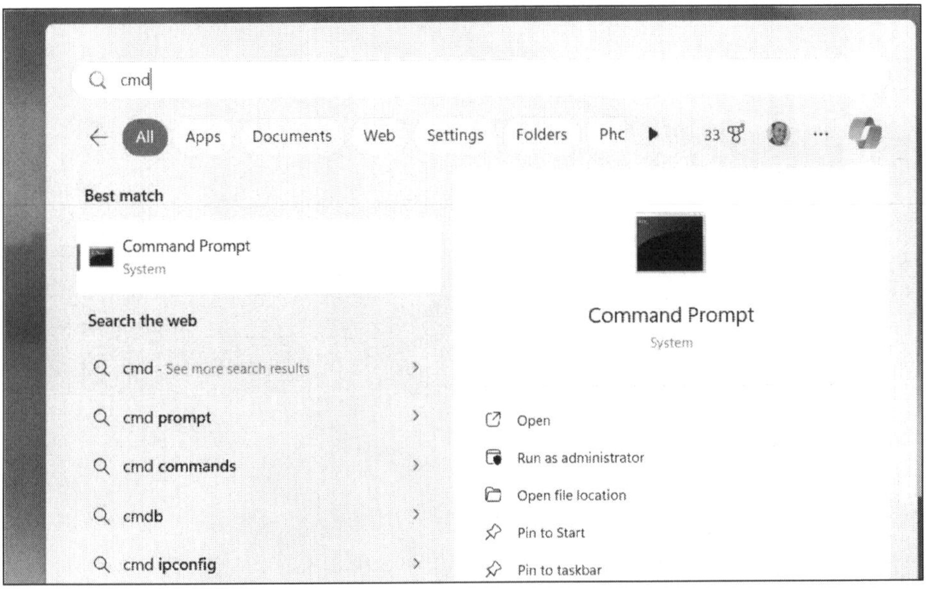

Check Python Version

- ☆ Type python —version and press Enter.
- ☆ You should see "Python 3.8.10" displayed, indicating that Python has been installed successfully.

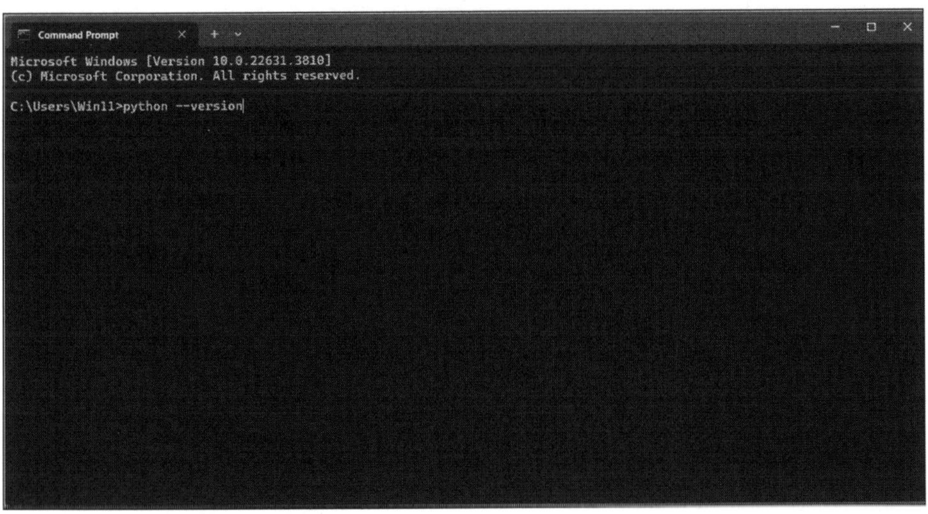

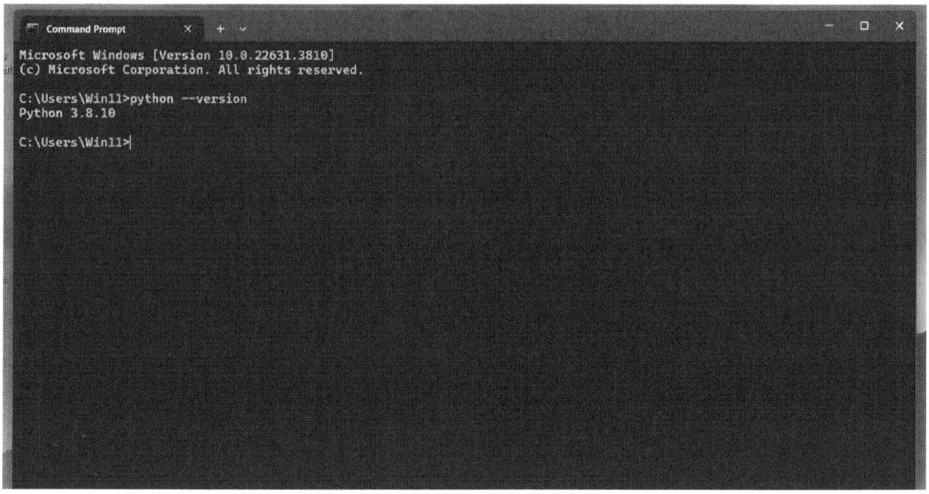

References

Viduka, D., Kraguljac, V. and Lièina, B., 2021. A COMPARATIVE ANALYSIS OF THE BENEFITS OF PYTHON AND JAVA FOR BEGINNERS. Quaestus, (19), pp.318-327.

Martelli, A., Ravenscroft, A.M., Holden, S. and McGuire, P., 2023. Python in a Nutshell. " O'Reilly Media, Inc.".

https: //www.python.org/downloads/

https: //discuss.python.org/t/python-installation/40571

Chapter 2

Basics and Installation of the PyCharm Platform

PyCharm is an integrated development environment (IDE) designed specifically for Python programming. Developed by JetBrains, PyCharm provides a comprehensive set of tools and features to streamline and enhance the software development process. It's suitable for both beginners and professional developers, offering a range of functionalities that make coding, debugging, and testing more efficient and enjoyable.

Key Features of PyCharm

1. **Code Editor**

   ☆ **Intelligent Code Completion:** Provides context-aware suggestions as you type, helping you write code faster and with fewer errors.

   ☆ **Syntax Highlighting:** Enhances code readability by coloring keywords, variables, and other elements differently.

   ☆ **Code Formatting:** Automatically formats your code according to PEP 8 standards or your own preferences.

2. **Project and Code Navigation**

   ☆ **Project Explorer:** Organize and manage your files and directories efficiently.

   ☆ **Go to Definition:** Quickly navigate to the definition of variables, classes, and functions.

- **Find Usages:** Locate where particular variables, methods, or classes are used throughout the project.

3. **Refactoring**
   - **Rename:** Safely rename variables, classes, methods, and more across your entire project.
   - **Extract Method/Variable:** Simplify your code by extracting sections into new methods or variables.
   - **Safe Delete:** Ensure that deleting code does not break any dependencies.

4. **Debugger**
   - **Breakpoints:** Set breakpoints to pause the execution of your code at specific lines.
   - **Step Through Code:** Step into, over, and out of functions to inspect how your code executes.
   - **Variable Inspection:** View the current state of variables and data structures during execution.

5. **Testing**
   - **Unit Testing:** Integrated support for popular testing frameworks like pytest and unittest.
   - **Test Runner:** Run tests and view results within the IDE.
   - **Code Coverage:** Analyze test coverage to ensure your code is thoroughly tested.

6. **Version Control**
   - **Integrated VCS:** Support for Git, SVN, Mercurial, and other version control systems.
   - **Commit and Push:** Commit changes, push to remote repositories, and manage branches directly from the IDE.
   - **Diff Viewer:** Compare different versions of files and view changes.

7. **Web Development**
   - **Framework Support:** Integrated tools for Django, Flask, and other web frameworks.
   - **HTML, CSS, and JavaScript:** Built-in editors for web technologies with code completion and validation.
   - **Live Editing:** Preview web pages and see changes in real time.

8. **Database Tools**
   - ☆ **Database Integration:** Connect to databases, run queries, and manage data directly within PyCharm.
   - ☆ **SQL Editor:** Write and execute SQL queries with syntax highlighting and code completion.
   - ☆ **Data Viewer:** View and edit database tables and records.

9. **Plugins and Customization**
   - ☆ **Plugin Marketplace:** Access a wide range of plugins to extend PyCharm's functionality.
   - ☆ **Themes:** Customize the appearance of the IDE with different themes and color schemes.
   - ☆ **Keymap:** Personalize keyboard shortcuts to match your workflow.

10. **Learning and Support**
    - ☆ **Integrated Tutorials:** Access built-in tutorials to get started with PyCharm.
    - ☆ **Documentation:** Comprehensive documentation covering all features and functionalities.
    - ☆ **Community:** A large and active user community provides additional support and resources.

Editions of PyCharm

1. **Community Edition**
   - ☆ **Free and Open Source:** Suitable for pure Python development.
   - ☆ **Basic Features:** Includes essential tools and functionalities.

2. **Professional Edition**
   - ☆ **Paid Version:** Offers a free trial with extended features.
   - ☆ **Advanced Tools:** Additional support for web development, scientific tools, and databases.
   - ☆ **Framework Support:** Enhanced support for Django, Flask, and other frameworks.

Why Use PyCharm?

- ☆ **Efficiency:** PyCharm's productivity features can significantly speed up your development process.
- ☆ **Reliability:** Built-in testing tools and analysis tools help ensure your code is robust and error-free.

☆ **Convenience:** Integrating all necessary tools into one environment reduces the need to switch between different applications.

☆ **Community and Support:** PyCharm has a large user base and active community, making it easier to find help and resources.

Overall, PyCharm is a powerful IDE that can greatly enhance your Python development experience, whether working on small scripts, large applications, or anything in between.

The Installation and Running of 1st Python Program

1. Open Web Browser

☆ Launch your preferred web browser (*e.g.*, Chrome, Edge, Firefox).

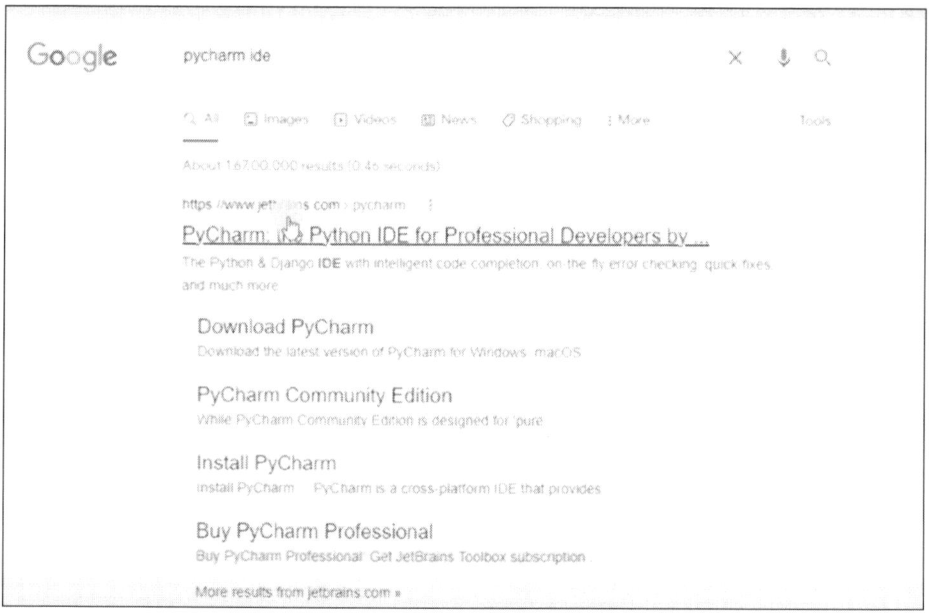

2. **Visit the Official Website**

   ☆ Go to the official PyCharm download page.

3. **Choose the Edition**

   ☆ You will see two editions: Community (free) and Professional (paid). Choose the one that suits your needs. For this guide, we'll proceed with the Community edition.

   ☆ Click on the "Download" button under the Community edition.

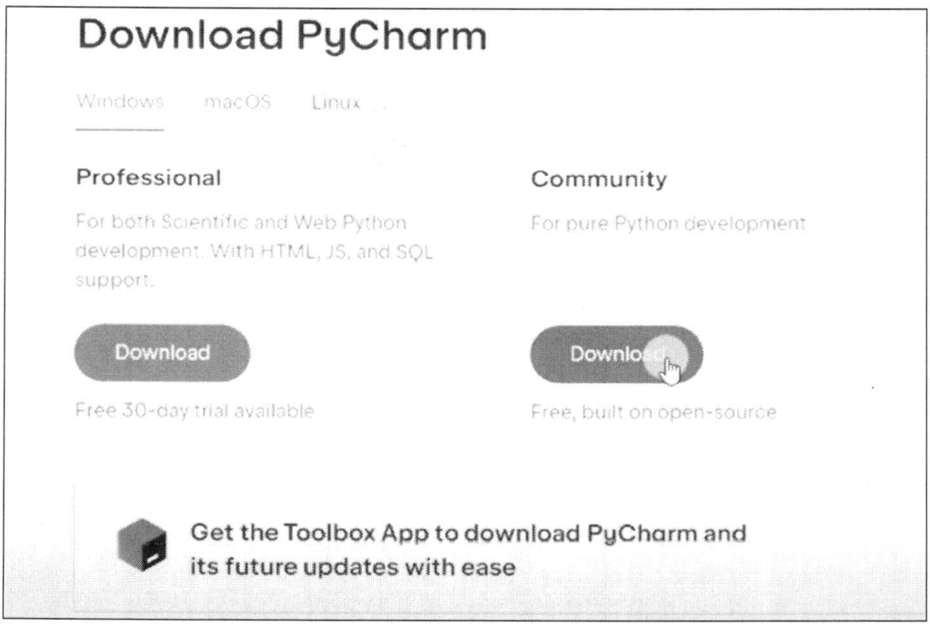

4. Locate the Downloaded Installer

- ✰ Once the download is complete, locate the downloaded.exe file (usually found in the "Downloads" folder).

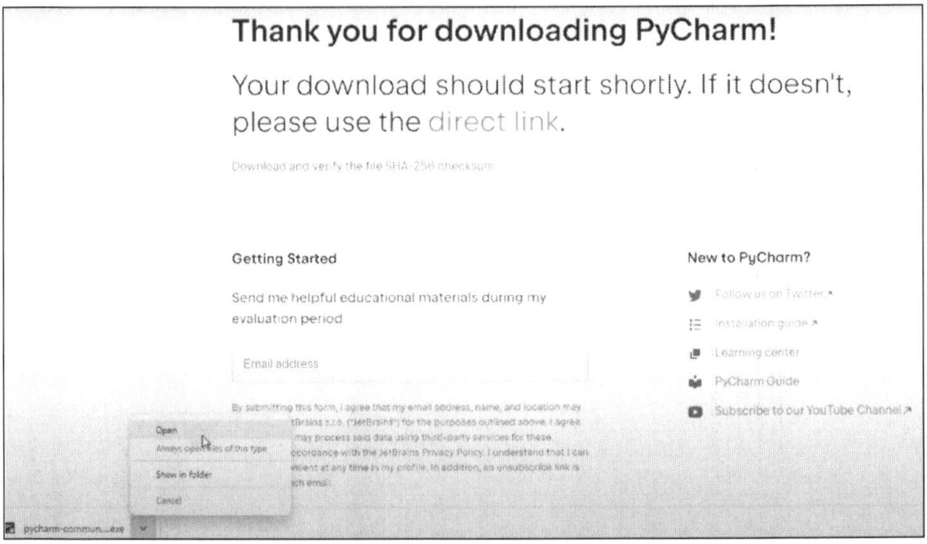

5. Run the Installer

- ✰ Double-click the.exe file to start the installation process.
- ✰ If prompted by User Account Control (UAC), click "Yes" to allow the installer to make changes to your device.

6. Start Installation

- ✰ The PyCharm Setup wizard will open. Click "Next" to start the installation process.

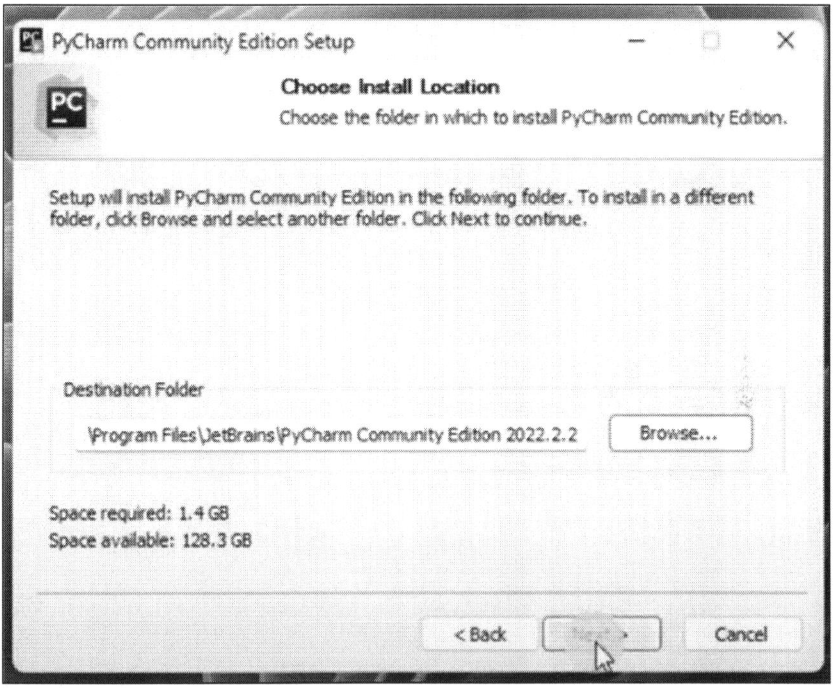

7. Choose Installation Location

- ☆ By default, PyCharm will be installed in

  `C: \Program Files\JetBrains\PyCharm Community Edition 2023.1.`

  You can change the installation path if needed. Click "Next" after choosing the location.

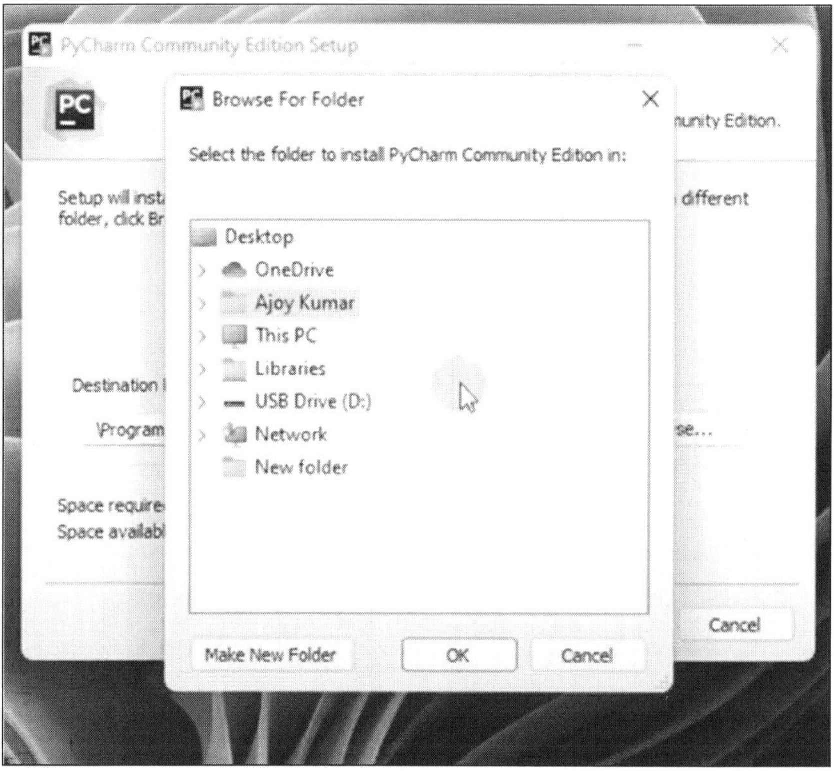

8. **Select Installation Options**

   - **Create Desktop Shortcut:** Check this box if you want a shortcut on your desktop.
   - **Add "Open Folder as Project" context menu:** Check this box to add an option to the right-click context menu for easy project access.
   - **Update PATH variable:** Check this to add PyCharm to your system PATH, making it easier to run from the command line.
   - **Create Associations:** Check the.py option to associate Python files with PyCharm.

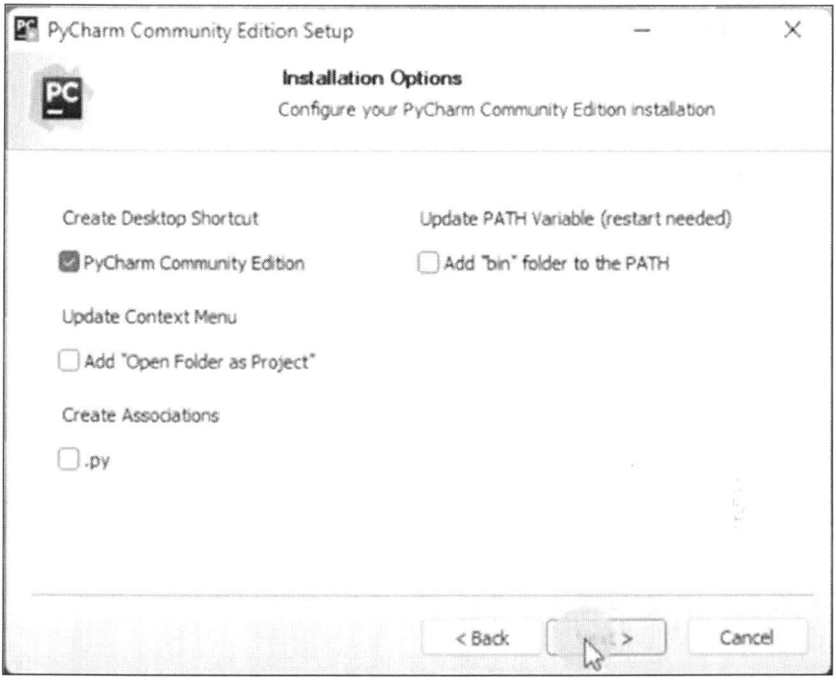

9. Choose Start Menu Folder

☆ You can select a folder name for PyCharm in the Start Menu or leave it as default. Click "Install" to begin the installation.

10. Installation Progress and Complete Installation

- ☆ Wait for the installation to complete. This might take a few minutes.
- ☆ After the installation, you will see a "Completing the PyCharm Setup Wizard" screen. Check "Run PyCharm Community Edition" if you want to launch it immediately, then click "Finish."

11. First Launch

- ✰ If you checked "Run PyCharm Community Edition" during installation, PyCharm will launch. Otherwise, double-click the PyCharm shortcut on your desktop or find it in the Start Menu.

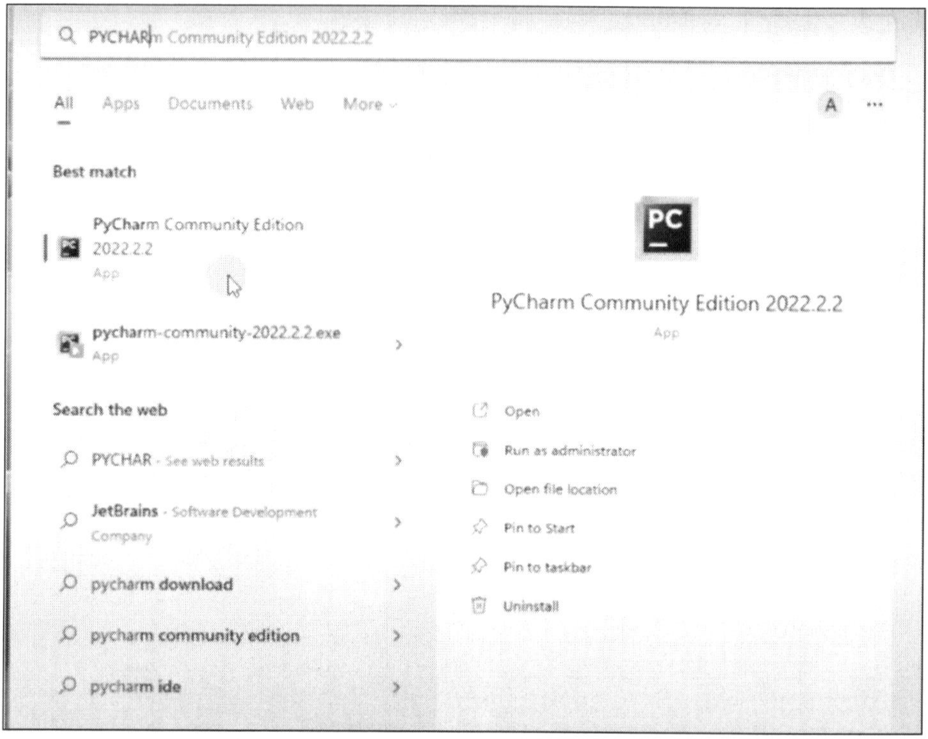

12. After PyCharm logo window appears then u can see an interphase of User Agreement and Data Sharing Agreement
    - ☆ Check the box and agree to continue
    - ☆ Check on Don't now

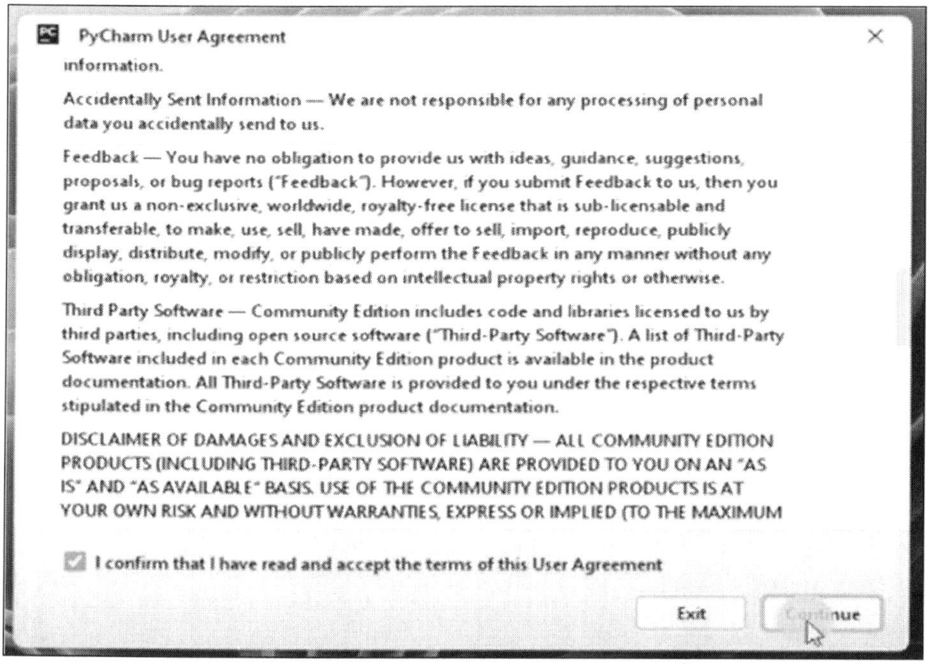

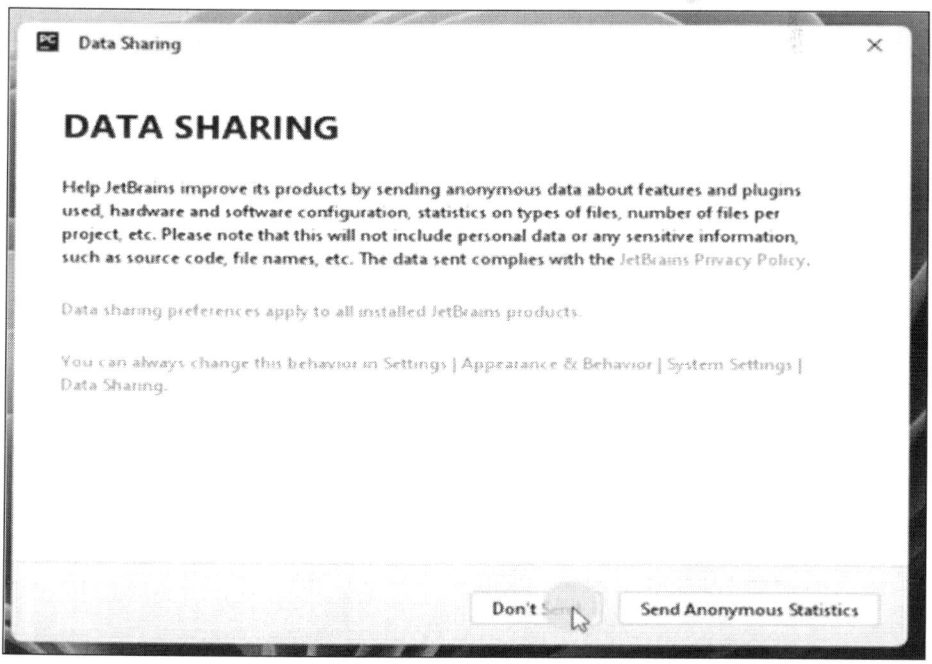

13. The interphase Showing New project would be showing, click on New Project

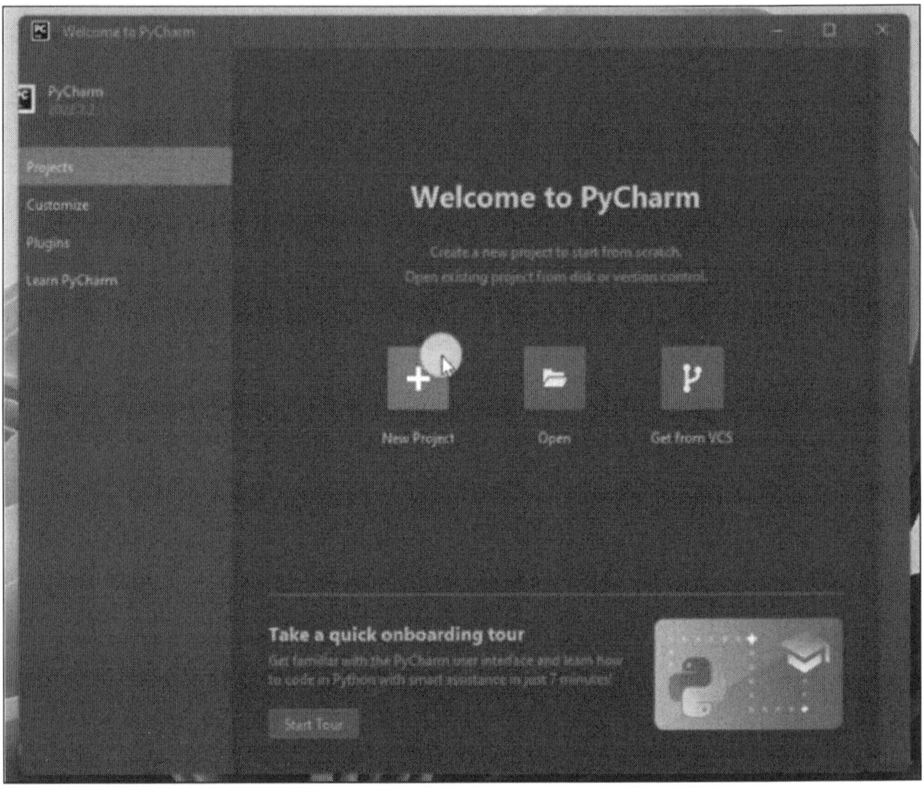

14. Settings for initiating the New Project

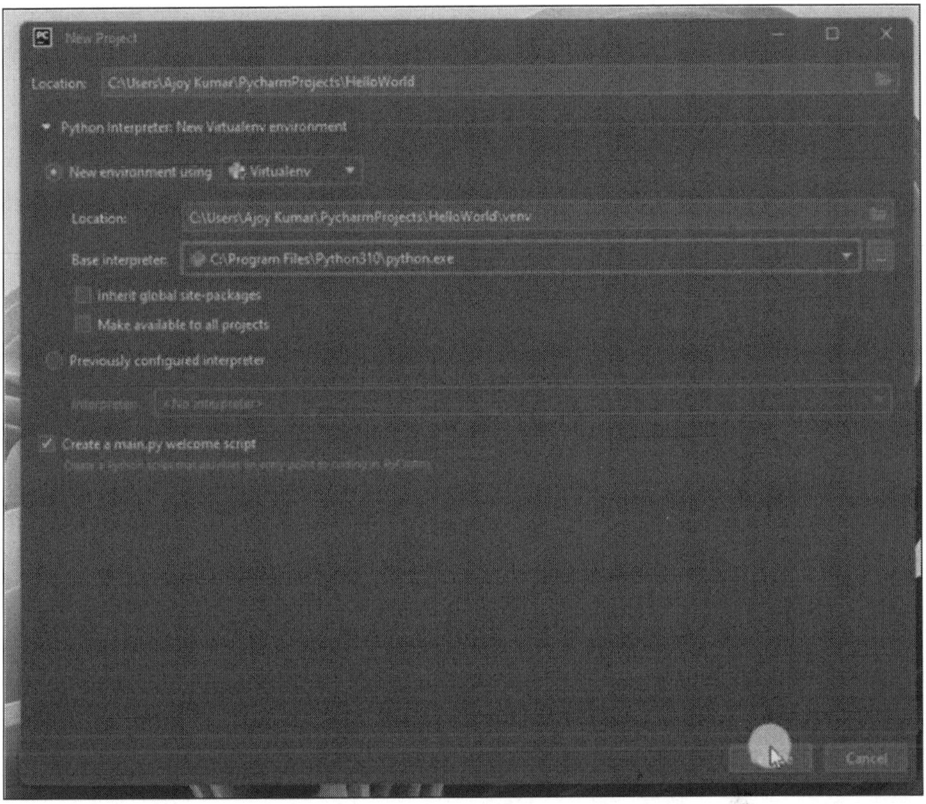

15. The Sample 1st program at Pycharm

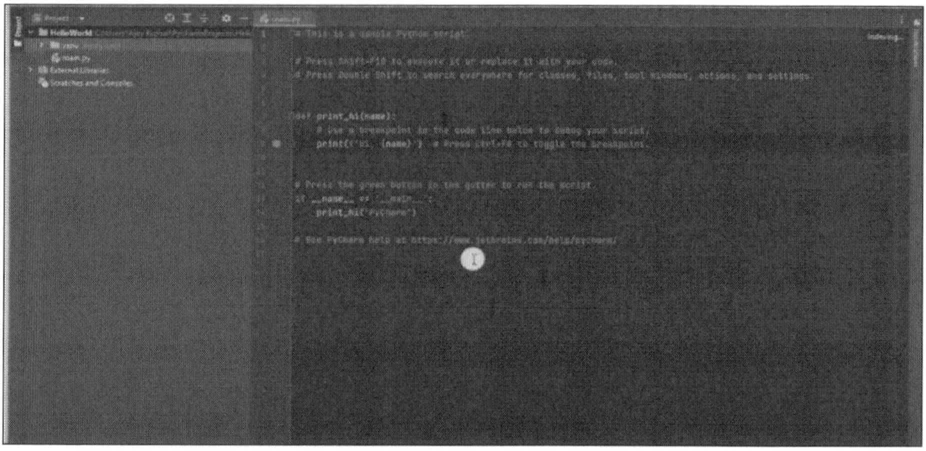

16. Run the Sample Program on PyCharm

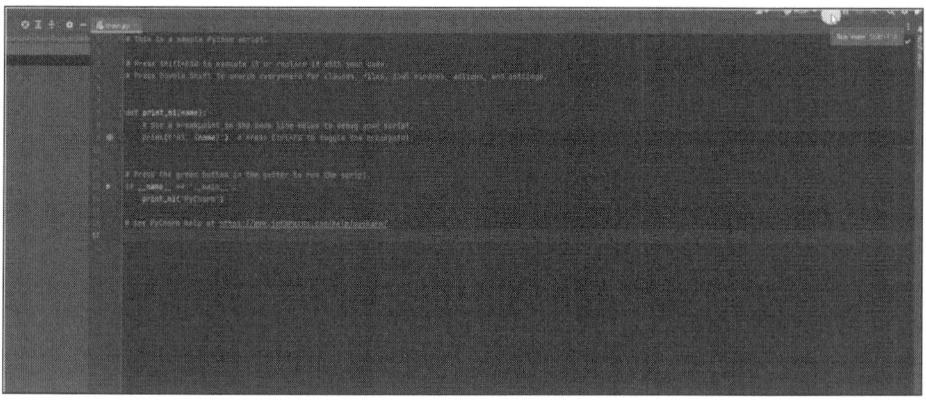

17. The terminal and Command Window

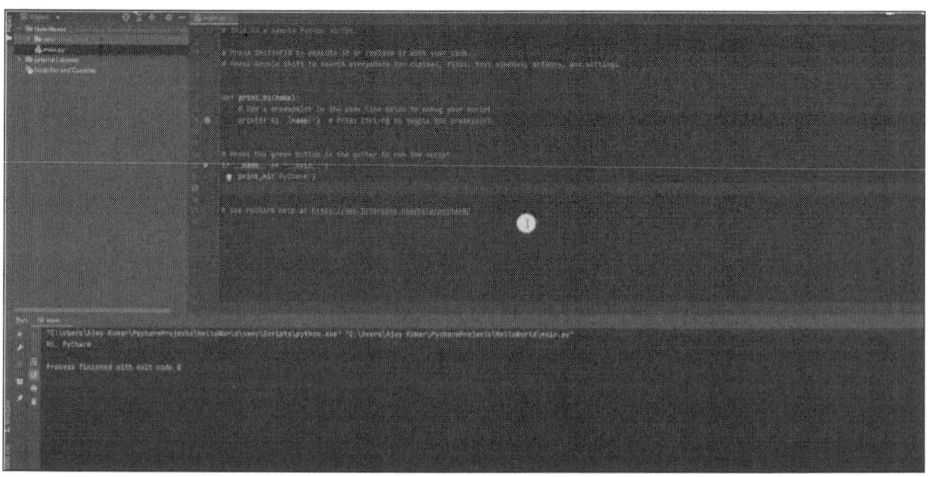

References

Islam, Q.N., 2015. *Mastering PyCharm*. Packt Publishing Ltd.

Nguyen, Q., 2019. *Hands-on application development with PyCharm: Accelerate your Python applications using practical coding techniques in PyCharm*. Packt Publishing Ltd.

Saabith, S., Vinothraj, T. and Fareez, M., 2021. A review on Python libraries and Ides for Data Science. *Int. J. Res. Eng. Sci*, *9*(11), pp.36-53.

https: //www.jetbrains.com/pycharm/download/

https: //www.jetbrains.com/help/pycharm/installation-guide.html

Chapter 3

Practical on Python Programming

Writing your First Python Program

You can edit and compile your written code directly. So, test yourself with **Python's first program.:**

```
print("Hello World! ")
```

Output:

Hello World!

Comments in Python

Comments in Python are the lines in the code that are ignored by the interpreter during the execution of the program. Also, Comments enhance the readability of the code and help the programmers to understand the code very carefully.

```
# sample comment
# This is Python Comment
name = "geeksforgeeks"
print(name)
```

Output:

Geeksforgeeks

Keywords in Python

Keywords in Python are reserved words that can not be used as a variable name, function name, or any other identifier.

and, as, for, or, def, is, in, try, return, not, except, continue, break, class, else, none, if, for, while, true, false, *etc.*

Python Variable

Python Variable is containers that store values. Python is not "statically typed". An Example of a Variable in Python is a representational name that serves as a pointer to an object. Once an object is assigned to a variable, it can be referred to by that name.

Rules for Python Variables

A Python variable name must start with a letter or the underscore character.

A Python variable name cannot start with a number.

A Python variable name can only contain alpha-numeric characters and underscores (A-z, 0-9, and _).

Variable in Python names are case-sensitive (name, Name, and NAME are three different variables).

The reserved words(keywords) in Python cannot be used to name the variable in Python.

Example

```
# An integer assignment
age = 45

# A floating point
salary = 1456.8

# A string
name = "John"

print(age)
print(salary)
print(name)
```

Output:

451456.8

John

Python Data Types

Data types are the classification or categorization of data items. It represents the kind of value that tells what operations can be performed on a particular data. Since everything is an object in Python programming, data types are classes and variables are instances (objects) of these classes.

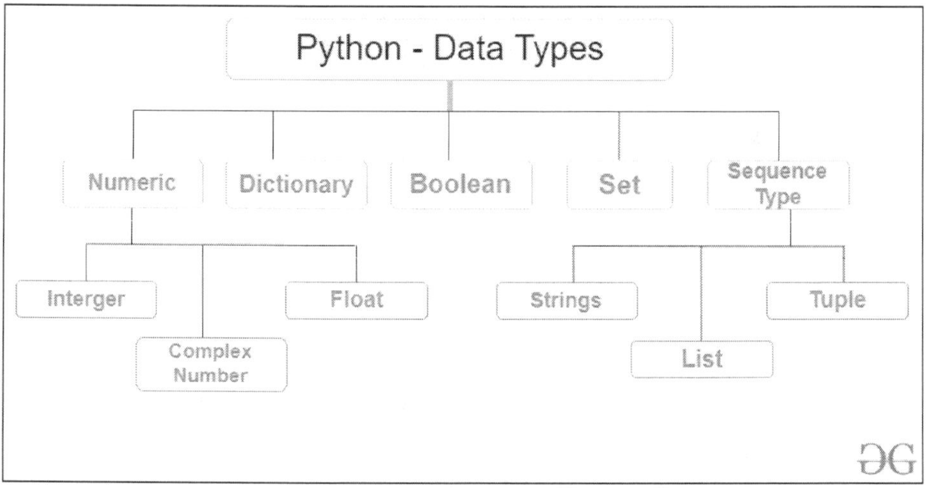

Example: This code assigns variable 'x' different values of various data types in Python.

```
x = "Hello World" # string
x = 50 # integer
x = 60.5 # float
x = 3j # complex
x = ["geeks", "for", "geeks"] # list
x = ("geeks", "for", "geeks") # tuple
x = {"name": "Suraj", "age": 24} # dict
x = {"geeks", "for", "geeks"} # set
x = True # bool
x = b"Geeks" # binary
```

Python Input/Output

This function first takes the input from the user and converts it into a string. The type of the returned object always will be <class 'str'>. It does not evaluate the expression it just returns the complete statement as String, and will print it.

```
# Python program show input and Output
val = input("Enter your value: ")
print(val)
```

```
Output:
Enter your value: 5
5
```

Python Operators

In Python programming, Operators in general are used to perform operations on values and variables. These are standard symbols used for the purpose of logical and arithmetic operations. In this article, we will look into different types of Python operators.

Arithmetic Operators

Python Arithmetic operators are used to perform basic mathematical operations like addition, subtraction, multiplication, and division.

Precedence of Arithmetic Operators

The precedence of Arithmetic Operators in Python is as follows:

P – Parentheses

E – Exponentiation

M – Multiplication (Multiplication and division have the same precedence)

D – Division

A – Addition (Addition and subtraction have the same precedence)

S – Subtraction

Example:

```
a = 9
b = 4
add = a + b
sub = a - b
mul = a * b
```

```
mod = a per cent b
p = a ** bprint(add)
print(sub)
print(mul)
print(mod)
print(p)

Output: 13
5
36
1
6561
```

Logical Operators

Python Logical operators perform Logical AND, Logical OR, and Logical NOT operations. It is used to combine conditional statements.

```
a = True
b = False
print(a and b)
print(a or b)
print(not a)

Output:
False
True
False
```

Bitwise Operators

Python Bitwise operators act on bits and perform bit-by-bit operations. These are used to operate on binary numbers.

```
a = 10
b = 4
print(a and b)
```

```
print(a | b)
print(~a)
print(a ^ b)
print(a >> 2)
print(a << 2)
```

```
Output
0
14
-11
14
2
40
```

Assignment Operators

Python Assignment operators are used to assign values to the variables.

```
a = 10
b = a
print(b)
b += a
print(b)
b -= a
print(b)
b *= a
print(b)
b <<= a
print(b)
```

```
Output
10
20
10
100
102400
```

Python If Else

The if statement alone tells us that if a condition is true it will execute a block of statements and if the condition is false it won't. But if we want to do something else if the condition is false, we can use the else statement with the if statement to execute a block of code when the if condition is false.

Example 1: Python IF-Else

```
i = 20
if (i < 15):
    print("i is smaller than 15")
    print("i'm in if Block")
else:
    print("i is greater than 15")
    print("i'm in else Block")
print("i'm not in if and not in else Block")
```

Output

```
i is greater than 15
i'm in else Block
i'm not in if and not in else Block
```

Example 2: Python if-elif-else ladder

```
i = 20
if (i == 10):
    print("i is 10")
elif (i == 15):
    print("i is 15")
elif (i == 20):
    print("i is 20")
else:
    print("i is not present")
```

Output

```
i is 20
```

Python for Loop

Python For loop is used for sequential traversal *i.e.* it is used for iterating over an iterable like String, Tuple, List, Set, or Dictionary. Here we will see a "for" loop in conjunction with the range() function to generate a sequence of numbers starting from 0, up to (but not including) 10, and with a step size of 2. For each number in the sequence, the loop prints its value using the print() function.

```
for i in range(0, 10, 2):
    print(i)
```

Output

0

2

4

6

8

Python While Loop

In this example, the condition for while will be True as long as the counter variable (count) is less than 3.

```
# Python program to illustrate while loop
count = 0
while (count < 3):
    count = count + 1
    print("Hello Geek")
```

Output

Hello Geek

Hello Geek

Hello Geek

Array in Python

An array is a data structure that can contain or hold a fixed number of elements that are of the same Python data type. An array is composed of an element and an index. Index in an array is the location where an element resides. All elements have their respective indices. Index of an array always starts with 0.

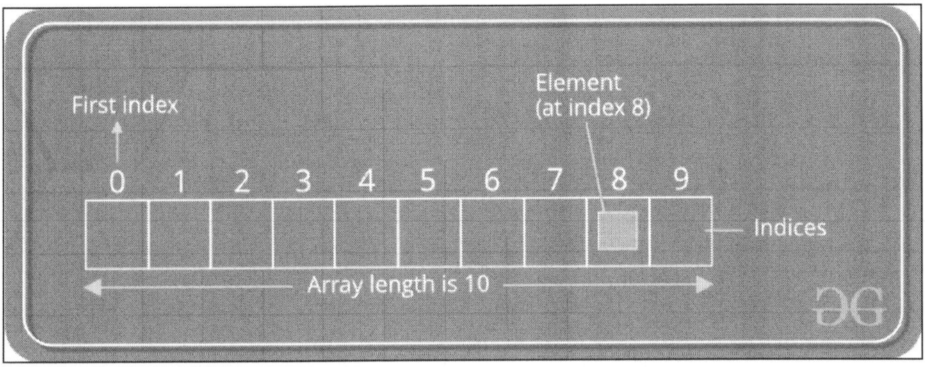


```
# Python program to illustrate Array
import array as arr
a = arr.array('d', [1.1, 3.5, 4.5])
print(a)
```

Output:

```
array('d', [1.1, 3.5, 4.5])
```

Class and Objects of Python

Python is an object-oriented programming language. This means that almost all the code is implemented using a special construct called classes. A class is a code template for creating objects.

Syntax: Class Definition

```
class ClassName:
# Statement
```

Syntax: Object Definition

```
obj = ClassName()
print(obj.attr)
```

```
# define a class
class Employee:
    # define a property
employee_id = 0
# create two objects of the Employee class
employee1 = Employee()
employee2 = Employee()
# access property using employee1
employee1.employeeID = 1001
print("Employee ID: {employee1.employeeID}")
# access properties using employee2
employee2.employeeID = 1002
print("Employee ID: {employee2.employeeID}")

Output:
Employee ID: 1001
Employee ID: 1002
```

Python Functions

Python Functions is a block of statements that return the specific task. The idea is to put some commonly or repeatedly done tasks together and make a function so that instead of writing the same code again and again for different inputs, we can do the function calls to reuse code contained in it over and over again.

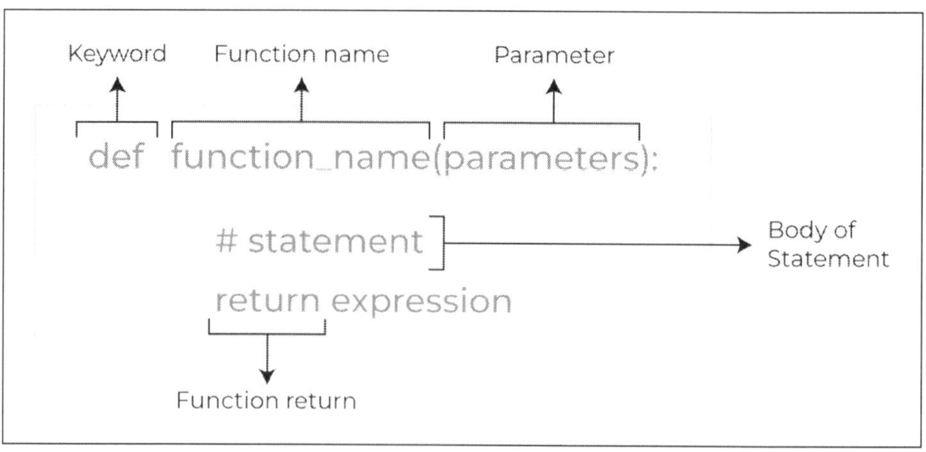

Note: There are mainly two types of functions in Python.

Built-in library function: These are Standard functions in Python that are available to use.

User-defined function: We can create our own functions based on our requirements.

Example 1:

```
# A simple Python function to check
# whether x is even or odd
def evenOdd(x):
    if (x per cent 2 == 0):
      print("even")
    else:
      print("odd")

# Driver code to call the function
evenOdd(2)
evenOdd(3)
```

Output:

even

odd

Example 2 : We return a value from the function using the return statement.

```
# function definitiondef find_square(num):
    result = num * num
    return result

# function call
square = find_square(3)
print('Square: ', square)
```

Output

Square: 9

References

Van Rossum, G., 2007, June. Python programming language. In *USENIX annual technical conference* (Vol. 41, No. 1, pp. 1-36).

Srinath, K.R., 2017. Python–the fastest growing programming language. *International Research Journal of Engineering and Technology*, *4*(12), pp.354-357.

VanderPlas, J., 2016. *Python data science handbook: Essential tools for working with data*. " O'Reilly Media, Inc.".

Drake, F.L. ed., 2003. *An introduction to Python* (p. 115). Bristol: Network Theory Ltd.

https: //www.geeksforgeeks.org/python-data-structures/

https: //docs.python.org/3/tutorial/datastructures.html

Chapter 4

Introduction to Artificial Intelligence, Machine Learning and Deep Learning

What is Artificial Intelligence?

Artificial Intelligence (AI) refers to the development of computer systems of performing tasks that require human intelligence. AI aids, in processing amounts of data identifying patterns and making decisions based on the collected information. This can be achieved through techniques like Machine Learning, Natural Language Processing, Computer Vision and Robotics. AI encompasses a range of abilities including learning, reasoning, perception, problem solving, data analysis and language comprehension. The ultimate goal of AI is to create machines that can emulate capabilities and carry out diverse tasks, with enhanced efficiency and precision. The field of AI holds potential to revolutionize aspects of our daily lives.

Examples of Artificial Intelligence

1. Virtual Personal Assistants: Popular examples like Siri, Google Assistant, and Amazon Alexa utilize AI to understand and respond to user commands. These assistants employ natural language processing (NLP) and machine learning algorithms to improve their accuracy and provide more personalized responses over time.
2. Autonomous Vehicles: AI powers the development of self-driving cars, trucks, and drones. Companies like Tesla, Waymo, and Uber are at the

forefront of this technology, using AI algorithms to analyse sensory data from cameras, radar, and lidar to make real-time driving decisions.

3. Healthcare Diagnosis and Treatment: AI algorithms are used to analyse medical data, including patient records, imaging scans, and genetic information, to assist healthcare professionals in diagnosing diseases and planning treatments. IBM's Watson for Health and Google's DeepMind are examples of AI platforms employed in healthcare.
4. Recommendation Systems: Online platforms like Netflix, Amazon, and Spotify utilize AI to analyse user behaviour and preferences, providing personalized recommendations for movies, products, and music. These systems employ collaborative filtering and content-based filtering techniques to enhance user experience and increase engagement.
5. Fraud Detection: AI algorithms are employed by financial institutions to detect fraudulent activities in real-time. These systems analyse.

Importance of Artificial Intelligence

Today, the amount of data in the world is so humongous that humans fall short of absorbing, interpreting, and making decisions of the entire data. This complex decision-making requires higher cognitive skills than human beings. This is why we're trying to build machines better than us, in these task. Another major characteristic that AI machines possess but we don't is repetitive learning. Let consider an example of how Artificial Intelligence is important to us. Data that is fed into the machines could be real-life incidents. How people interact, behave and react *etc.*? So, in other words, machines learn to think like humans, by observing and learning from humans. That's precisely what is called Machine Learning which is a subfield of AI. Humans are observed to find repetitive tasks highly boring. Accuracy is another factor in which we humans lack. Machines have extremely high accuracy in the tasks that they perform. Machines can also take risks instead of human beings.

How Does AI Work ?

Artificial Intelligence (AI) uses a wide range of techniques and approaches that enable machines to simulate human-like intelligence and perform tasks that traditionally require human assistance. AI systems work through a combination of algorithms, data, and computational power. Here's an overview of how AI works:

1. Data Collection: AI systems rely on vast amounts of data to learn and make decisions. Data can be collected from various sources, including sensors, digital devices, databases, the internet, and user interactions. The quality and quantity of data are crucial for training accurate and reliable AI models.
2. Data Pre-processing: Once data is collected, it needs to be pre-processed to ensure it's clean, structured, and suitable for analysis. This pre-processing

stage may involve tasks such as cleaning noisy data, handling missing values, standardizing formats, and encoding categorical variables.

3. Algorithm Selection: AI algorithms are chosen based on the specific task or problem the AI system aims to solve. Different algorithms are suited for different types of tasks, such as classification, regression, clustering, and pattern recognition. Common AI algorithms include neural networks, decision trees, support vector machines, and k-nearest neighbours.
4. Model Training: In the training phase, AI models are fed with labelled data (supervised learning) or unlabelled data (unsupervised learning) to learn patterns and relationships. During training, the model adjusts its parameters iteratively to minimize errors and improve its performance on the given task. This process involves optimization techniques like gradient descent and backpropagation in neural networks.
5. Model Evaluation: After training, the AI model is evaluated using separate validation data to assess its performance and generalization ability. Performance metrics such as accuracy, precision, recall, F1-score, and area under the curve (AUC) are used to quantify the model's effectiveness in making predictions or decisions.
6. Model Deployment: Once the AI model meets the desired performance criteria, it can be deployed into production environments to perform real-world tasks. Deployment involves integrating the model into existing systems, such as mobile apps, web services, or embedded devices, to provide AI-driven functionalities.
7. Continuous Learning and Improvement: AI systems can adapt and improve over time through continuous learning. They can be updated with new data and retrained periodically to stay relevant and accurate in dynamic environments. Techniques like online learning, transfer learning, and reinforcement learning enable AI models to learn from new experiences and feedback.
8. Inference and Decision-Making: During inference, the trained AI model applies its learned knowledge to make predictions or decisions on new, unseen data. Inference involves feeding input data into the model and obtaining output predictions or classifications based on the model's learned patterns and representations.

What are the Applications of AI?

Artificial Intelligence (AI) has a wide range of applications and has been adopted in many industries to improve efficiency, accuracy, and productivity. Some of the most common uses of AI are:

1. Healthcare: AI is used in healthcare for various purposes such as diagnosing diseases, predicting patient outcomes, drug discovery, and personalized treatment plans.

2. Finance: AI is used in the finance industry for tasks such as credit scoring, fraud detection, portfolio management, and financial forecasting.
3. Retail: AI is used in the retail industry for applications such as customer service, demand forecasting, and personalized marketing.
4. Manufacturing: AI is used in manufacturing for tasks such as quality control, predictive maintenance, and supply chain optimization.
5. Transportation: AI is used in transportation for optimizing routes, improving traffic flow, and reducing fuel consumption.
6. Education: AI is used in education for personalizing learning experiences, improving student engagement, and providing educational resources.
7. Marketing: AI is used in marketing for tasks such as customer segmentation, personalized recommendations, and real-time audience analysis.
8. Gaming: AI is used in gaming for developing intelligent game characters and providing personalized gaming experiences.
9. Security: AI is used in security for tasks such as facial recognition, intrusion detection, and cyber threat analysis.
10. Natural Language Processing (NLP): AI is used in NLP for tasks such as speech recognition, machine translation, and sentiment analysis.

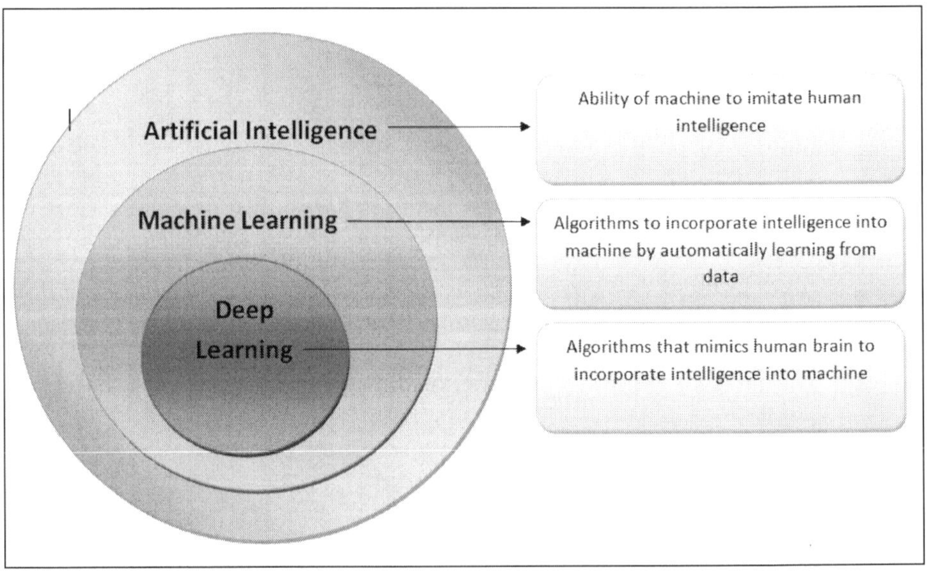

Machine Learning

Introduction

Machine learning is a branch of artificial intelligence (AI) and computer science which focuses on the use of data and algorithms to imitate the way that humans learn, gradually improving its accuracy. Machine learning research enables machines to learn new abilities, gain new knowledge, and reorganize previously acquired knowledge Machine learning is an important component of the growing field of data science. Through the use of statistical methods, algorithms are trained to make classifications or predictions, and to uncover key insights in data mining projects. These insights subsequently drive decision making within applications and businesses, ideally impacting key growth metrics. As big data continues to expand and grow, the market demand for data scientists will increase. They will be required to help identify the most relevant business questions and the data to answer them. Classical machine learning is often categorized by how an algorithm learns to become more accurate in its predictions.

There are four basic approaches: supervised learning, unsupervised learning, semi-supervised learning and reinforcement learning. The type of algorithm data scientists choose to use depends on what type of data they want to predict. Decision tree, Naive Bayes, Support vector machine, Artificial neural network, K nearest neighbour, Deep learning, Ensemble learning are some examples of machine learning

Why Machine Learning?:

Most problems in Remote Sensing require 'detecting' objects or land cover types from images. Earth Observation data is noisy. Hard to determine rules that can separate one class from the other. Building models require expert knowledge and can be tedious. Machine learning algorithms takes the guesswork out of determining thresholds and cut-offs. Machine learning is growing in importance due to increasingly enormous volumes and variety of data, the access and affordability of computational power, and the availability of high speed Internet. These digital transformation factors make it possible for one to rapidly and automatically develop models that can quickly and accurately analyze extraordinarily large and complex data sets. There are a multitude of use cases that machine learning can be applied to in order to cut costs, mitigate risks, and improve overall quality of life including recommending products/services, detecting cybersecurity breaches, and enabling self-driving cars. With greater access to data and computation power, machine learning is becoming more ubiquitous every day and will soon be integrated into many facets of human life.

Deep Learning Algorithms

What is Deep Learning Algorithm?

Deep learning can be defined as the method of machine learning and artificial intelligence that is intended to intimidate humans and their actions based on certain human brain functions to make effective decisions. It is a very important data science element that channels its modeling based on data-driven techniques under predictive modeling and statistics. To drive such a human-like ability to adapt and learn and to function accordingly, there have to be some strong forces which we popularly called algorithms.

Deep learning algorithms are dynamically made to run through several layers of neural networks, which are nothing but a set of decision-making networks that are pre-trained to serve a task. Later, each of these is passed through simple layered representations and move on to the next layer. However, most machine learning is trained to work fairly well on datasets that have to deal with hundreds of features or columns. For a data set to be structured or unstructured, machine learning tends to fail mostly because they fail to recognize a simple image having a dimension of 800x1000 in RGB. It becomes quite unfeasible for a traditional machine learning algorithm to handle such depths. This is where deep learning.

Importance of Deep Learning

Deep learning algorithms play a crucial role in determining the features and can handle the large number of processes for the data that might be structured or unstructured. Although, deep learning algorithms can overkill some tasks that might involve complex problems because they need access to huge amounts of data so that they can function effectively. For example, there's a popular deep learning tool that recognizes images namely Imagenet that has access to 14 million images in its dataset-driven algorithms. It is a highly comprehensive tool that has defined a next-level benchmark for deep learning tools that aim images as their dataset.

Deep learning algorithms are highly progressive algorithms that learn about the image that we discussed previously by passing it through each neural network layer. The layers are highly sensitive to detect low-level features of the image like edges and pixels and henceforth the combined layers take this information and form holistic representations by comparing it with previous data. For example, the middle layer might be programmed to detect some special parts of the object in the photograph which other deep trained layers are programmed to detect special objects like dogs, trees, utensils, *etc.*

However, if we talk out the simple task that involves less complexity and a data-driven resource, deep learning algorithms fail to generalize simple data. This is one of the main reasons deep learning is not considered effective as linear or boosted tree models. Simple models aim to churn out custom data, track fraudulent transactions and deal with less complex datasets with fewer features. Also, there are various cases like multiclass classification where deep learning can be effective because it involves smaller but more structured datasets but is not preferred usually.

Deep Learning is a part of Machine Learning used to solve complex problems and build intelligent solutions. The core concept of Deep Learning has been derived from the structure and function of the human brain. Deep Learning uses artificial neural networks to analyze data and make predictions. It has found its application in almost every sector of business. In this article, we'll learn the top Deep Learning

Deep Learning Applications

1. Virtual Assistants

Virtual Assistants are cloud-based applications that understand natural language voice commands and complete tasks for the user. Amazon Alexa, Cortana, Siri, and Google Assistant are typical examples of virtual assistants. They need internet-connected devices to work with their full capabilities. Each time a command is fed to the assistant, they tend to provide a better user experience based on past experiences using Deep Learning algorithms.

2. Chatbots

Chatbots can solve customer problems in seconds. A chatbot is an AI application to chat online via text or text-to-speech. It is capable of communicating and performing actions similar to a human. Chatbots are used a lot in customer interaction, marketing on social network sites, and instant messaging the client. It delivers automated responses to user inputs. It uses machine learning and deep learning algorithms to generate different types of reactions.

3 Healthcare

Deep Learning has found its application in the Healthcare sector. Computer-aided disease detection and computer-aided diagnosis have been possible using Deep Learning. It is widely used for medical research, drug discovery, and diagnosis of life-threatening diseases such as cancer and diabetic retinopathy through the process of medical imaging.

4. Entertainment

Companies such as Netflix, Amazon, YouTube, and Spotify give relevant movies, songs, and video recommendations to enhance their customer experience. This is all thanks to Deep Learning. Based on a person's browsing history, interest, and behavior, online streaming companies give suggestions to help them make product and service choices. Deep learning techniques are also used to add sound to silent movies and generate subtitles automatically.

5. Robotics

Deep Learning is heavily used for building robots to perform human-like tasks. Robots powered by Deep Learning use real-time updates to sense obstacles in their path and pre-plan their journey instantly. It can be used to carry goods in hospitals, factories, warehouses, inventory management, manufacturing products, *etc.*

Boston Dynamics robots react to people when someone pushes them around, they can unload a dishwasher, get up when they fall, and do other tasks as well.

6. Natural Language Processing

Another important field where Deep Learning is showing promising results is NLP, or Natural Language Processing. It is the procedure for allowing robots to study and comprehend human language. However, keep in mind that human language is excruciatingly difficult for robots to understand. Machines are discouraged from correctly comprehending or creating human language not only because of the alphabet and words, but also because of context, accents, handwriting, and other factors.

References

Alpaydin, E., 2020. *Introduction to machine learning*. MIT press.

Campesato, O., 2020. *Artificial intelligence, machine learning, and deep learning*. Mercury Learning and Information.

Goodfellow, I., Bengio, Y. and Courville, A., 2016. *Deep learning*. MIT press.

Lauzon, F.Q., 2012, July. An introduction to deep learning. In *2012 11th international conference on information science, signal processing and their applications* (*ISSPA*) (pp. 1438-1439). IEEE.

Sajja, P.S. and Sajja, P.S., 2021. Introduction to artificial intelligence. *Illustrated Computational Intelligence: Examples and Applications*, pp.1-25.

Smola, A. and Vishwanathan, S.V.N., 2008. Introduction to machine learning. *Cambridge University, UK, 32*(34), p.2008.

Chatper 5

Introduction to Various Fields of Artificial Intelligence and Generative AI

Artificial intelligence (AI) is transforming our interaction with technology and reshaping numerous industries. Fundamentally, AI involves creating computer systems capable of performing tasks that typically require human intelligence, such as problem-solving, decision-making, understanding language, and visual perception. To truly comprehend what AI is, it's important to recognize that it includes a variety of technologies like machine learning, neural networks, natural language processing, and robotics.

Understanding AI highlights its significant impact on our daily lives and its potential to foster innovation across various sectors, including healthcare, finance, and transportation. From smart assistants and automated customer service to sophisticated data analysis, AI is leading the way toward a more intelligent and efficient future.

Artificial intelligence (AI) is a domain within computer science that focuses on designing systems capable of executing tasks traditionally requiring human intelligence. The definition of AI involves creating algorithms and models that allow machines to learn from data, identify patterns, and make decisions. This domain includes various subfields, such as machine learning, which involves systems improving their performance based on experience, and natural language processing, which enables machines to interpret and generate human language.

Comprehending AI extends beyond technical definitions to encompass its applications and implications. AI is integrated into everyday tools, such as virtual assistants, which use AI to process and respond to user inputs. Additionally, AI is enhancing efficiency across industries, including healthcare, with diagnostic tools, and finance, with fraud detection systems. Therefore, the definition and meaning of AI highlight its crucial role in technological advancement and operational improvement.

Artificial Intelligence (AI) is a multifaceted domain within computer science that encompasses a variety of subfields, each focusing on different aspects of enabling machines to perform tasks that typically require human intelligence. This chapter explores the primary fields within AI, highlighting their unique contributions and applications.

Machine Learning (ML)

Machine Learning (ML) is a cornerstone of AI, involving the creation of algorithms that enable systems to learn from and make decisions based on data. ML is divided into several categories:

1. Supervised Learning: In this approach, systems are trained on labeled data, learning to predict outcomes or classify data based on input examples. Applications include spam detection, image classification, and predictive maintenance.
2. Unsupervised Learning: Unlike supervised learning, this method involves analyzing and clustering unlabeled data to discover hidden patterns or intrinsic structures. Common applications include customer segmentation, anomaly detection, and market basket analysis.
3. Reinforcement Learning: This area focuses on training agents to make a sequence of decisions by rewarding desired behaviors and penalizing undesired ones. It's widely used in robotics, gaming, and autonomous vehicles.

Natural Language Processing (NLP)

Natural Language Processing (NLP) enables machines to understand, interpret, and generate human language. Key applications of NLP include:

1. Speech Recognition: Converting spoken language into written text, which is essential for virtual assistants and transcription services.
2. Sentiment Analysis: Identifying and extracting subjective information from text, such as opinions and emotions, often used in social media monitoring and customer feedback analysis.
3. Machine Translation: Automatically translating text from one language to another, exemplified by tools like Google Translate.
4. Text Generation: Creating human-like text based on given prompts or datasets, used in content creation and chatbots.

Computer Vision

Computer Vision focuses on enabling machines to interpret and understand visual information from the world. Key areas include:

1. Image Recognition: Identifying objects, people, or scenes within images, applied in areas such as security surveillance and photo tagging.
2. Facial Recognition: Verifying or identifying individuals from images or video frames, used in security systems and personalized user experiences.
3. Object Detection: Locating and identifying multiple objects within an image or video, essential for applications like autonomous driving and medical imaging.

Robotics

Robotics combines AI with engineering to create machines that can perform tasks autonomously. Notable applications include:

1. Autonomous Vehicles: Self-driving cars and drones that navigate and operate without human intervention.
2. Industrial Robots: Automating repetitive tasks in manufacturing and assembly lines, enhancing efficiency and precision.
3. Service Robots: Assisting in tasks such as cleaning, delivery, and customer service in various environments.

Expert Systems

Expert Systems are designed to emulate the decision-making abilities of a human expert. They consist of:

1. Rule-Based Systems: Utilizing a set of predefined rules to make decisions or solve problems, often used in diagnostics and troubleshooting.
2. Decision Support Systems: Helping users make informed decisions by analyzing data and presenting possible solutions, commonly used in business and healthcare.

Neural Networks

Neural Networks, inspired by the human brain, are a class of models within machine learning. They include:

1. Deep Learning: Utilizing multi-layered neural networks to model complex patterns in large datasets, used in image and speech recognition.
2. Convolutional Neural Networks (CNNs): Specialized in processing grid-like data such as images, widely used in visual recognition tasks.
3. Recurrent Neural Networks (RNNs): Designed for sequential data, like time series or text, and used in language modeling and translation.

Fuzzy Logic

Fuzzy Logic deals with reasoning that is approximate rather than fixed and exact. It's useful for handling uncertainty and imprecision, with applications in control systems and decision-making processes.

Genetic Algorithms

Genetic Algorithms are optimization techniques inspired by natural selection. They are part of evolutionary computation and are used to solve complex problems by iteratively improving candidate solutions, often applied in scheduling, engineering design, and machine learning model optimization.

Recommender Systems

Recommender Systems suggest products or services to users based on various algorithms:

1. Collaborative Filtering: Recommending items based on the preferences and behaviors of similar users, widely used in e-commerce and streaming services.
2. Content-Based Filtering: Recommending items based on the characteristics of items a user has shown interest in, used in personalized content delivery.

Generative AI

Generative AI is a cutting-edge branch of artificial intelligence that focuses on creating new content, such as images, text, music, and other data forms, by learning patterns from existing datasets. Unlike traditional AI, which is primarily used for tasks like classification, prediction, and decision-making, generative AI aims to produce novel outputs that are similar to the training data but not exact replicas. This capability is achieved through sophisticated machine learning techniques and models, including Generative Adversarial Networks (GANs), Variational Autoencoders (VAEs), and Transformers.

Key Technologies in Generative AI

1. Generative Adversarial Networks (GANs)

GANs are a class of AI models composed of two neural networks: the generator and the discriminator. The generator creates new data samples, while the discriminator evaluates their authenticity. The two networks are trained simultaneously in a process where the generator tries to produce increasingly realistic data, and the discriminator becomes better at distinguishing between real and generated data. This adversarial training process enhances the quality of the generated content over time.

2. Variational Autoencoders (VAEs)

VAEs are another type of generative model that learns to encode input data into a lower-dimensional latent representation and then decodes it to reconstruct the original data. By sampling from the learned latent space, VAEs can generate new data that shares similar characteristics with the training data. This approach is particularly useful for tasks like image generation and data compression.

3. Transformers

Transformers, especially in the context of natural language processing, have revolutionized the generation of human-like text. Models such as the Generative Pre-trained Transformer (GPT) predict the next word in a sentence based on the context provided by preceding words. This allows for the creation of coherent and contextually relevant text, enabling applications ranging from chatbots to automated content creation.

4. Diffusion Models

Diffusion models are a class of generative models that have recently gained attention for their ability to produce high-quality synthetic data, such as images, by modeling the process of data generation as a series of gradual transformations. These models are inspired by the concept of diffusion processes in physics, where particles spread from regions of high concentration to low concentration over time.

Applications of Generative AI in Fisheries

Generative AI has emerged as a groundbreaking technology with the potential to transform various industries, including fisheries. By harnessing the power of machine learning and advanced modeling techniques, generative AI can provide innovative solutions for generating fish population data, predicting fish behavior and migration, and optimizing fishing practices. This chapter delves into these applications, highlighting how generative AI can contribute to a more sustainable and efficient fisheries industry.

1. Fish Population Data Generation

- ☆ Enhancing Research and Management with Synthetic Data: One of the most significant challenges in fisheries management is the scarcity of comprehensive and diverse data on fish populations. Generative AI can address this issue by creating synthetic datasets that simulate various environmental and ecological scenarios. These synthetic datasets can be invaluable for researchers and policymakers, allowing them to study potential outcomes without relying solely on limited real-world data. For example, AI-generated data can model the impacts of climate change on fish populations, providing insights into how different species might respond to shifts in water temperature, salinity, and food availability.
- ☆ Modeling Population Dynamics: Generative AI models can simulate the complex dynamics of fish populations over time. By incorporating factors

such as breeding cycles, predation rates, and habitat changes, these models can predict how fish populations might fluctuate under different conditions. This predictive capability is crucial for proactive fisheries management, enabling the anticipation of population trends and the implementation of measures to ensure the sustainability of fish stocks. Moreover, these models can be used to explore the potential consequences of human activities, such as overfishing and habitat destruction, thereby informing conservation strategies.

2. Predicting Fish Behavior and Migration

- ☆ Behavioral Analysis and Simulation: Understanding fish behavior is essential for effective fisheries management and conservation. Generative AI can analyze historical data on fish behavior to identify patterns and predict future behaviors. For instance, AI models can predict spawning times, feeding habits, and social interactions among fish species. By generating realistic behavior simulations, AI helps researchers and managers understand how different environmental factors, such as temperature changes and food availability, influence fish behavior. These insights are vital for designing management practices that align with natural fish behaviors, minimizing disruptions to ecosystems.
- ☆ Migration Pattern Prediction: Fish migration is a critical aspect of their lifecycle, affecting reproduction, feeding, and survival. Generative AI can significantly enhance the understanding and prediction of fish migration patterns. By analyzing data from tracking devices, satellite imagery, and environmental sensors, AI models can predict the timing and routes of fish migrations. This capability is invaluable for fisheries management, as it allows for the implementation of measures to protect migratory pathways and ensure the sustainability of fish populations. Additionally, predicting migration patterns can help optimize the placement of marine protected areas and guide conservation efforts to safeguard critical habitats.

3. Optimizing Fishing Practices

- ☆ Promoting Sustainable Fishing: Generative AI can play a pivotal role in promoting sustainable fishing practices. By analyzing environmental data and fish behavior patterns, AI models can predict the optimal times and locations for fishing, minimizing overfishing and reducing bycatch. This ensures that fishing activities are conducted in a manner that maintains healthy fish populations and preserves marine biodiversity. Furthermore, generative AI can adapt fishing strategies in real-time, responding to changing environmental conditions and fish behaviors to maximize efficiency and sustainability.
- ☆ Reducing Environmental Impact: One of the significant challenges in fisheries is the unintended capture of non-target species, known as

bycatch. Generative AI can help design more selective fishing gear and techniques that target specific species, reducing bycatch and minimizing the environmental impact of fishing activities. For example, AI can be used to develop fishing nets that allow non-target species to escape, or to design bait that is more attractive to target species. Additionally, generative AI can assist in dynamic quota management, adjusting fishing quotas based on real-time data and predictive models to reflect current population dynamics and environmental conditions.

4. Enhancing Aquaculture Practices

- ✰ Optimizing Growth Conditions: Aquaculture, or fish farming, is a growing sector that can benefit immensely from generative AI. AI models can simulate optimal environmental conditions for fish farming, such as water quality, temperature, and feeding schedules, enhancing growth rates and overall health. By predicting the likelihood of disease outbreaks, AI can help aquaculture managers take preventive measures, reducing losses and improving productivity. For instance, AI can analyze water quality data and fish behavior to detect early signs of disease, enabling timely intervention.
- ✰ Improving Feed Efficiency: Feed represents a significant cost in aquaculture, and optimizing feed efficiency is crucial for profitability. Generative AI can assist in formulating optimal feed compositions based on the nutritional needs of different fish species, improving feed conversion rates and reducing waste. Additionally, AI can optimize feeding schedules to ensure fish are fed at the most beneficial times, promoting better growth and health. By continuously monitoring and adjusting feeding practices, generative AI helps maximize the efficiency and sustainability of aquaculture operations.

Ethical Considerations and Challenges

While generative AI holds immense potential, it also raises several ethical considerations and challenges:

1. Misinformation and Deepfakes

The ability to generate highly realistic images, videos, and text can be misused to create deepfakes and spread misinformation. This poses significant challenges for verifying the authenticity of digital content and combating the potential misuse of AI-generated media.

2. Intellectual Property

As AI models generate new content based on existing data, questions arise regarding intellectual property rights and the ownership of AI-generated creations. This area of law and ethics is still evolving, and clear guidelines are needed to address these issues.

3. Bias and Fairness

Generative models can inadvertently learn and propagate biases present in their training data. Ensuring that generative AI systems produce fair and unbiased content requires careful consideration of the data used for training and the implementation of techniques to mitigate bias.

References

Ertel, W., 2018. *Introduction to artificial intelligence*. Springer.

Feuerriegel, S., Hartmann, J., Janiesch, C. and Zschech, P., 2024. Generative ai. *Business and Information Systems Engineering*, *66*(1), pp.111-126.

Jo, A., 2023. The promise and peril of generative AI. *Nature*, *614*(1), pp.214-216.

Xu, Y., Liu, X., Cao, X., Huang, C., Liu, E., Qian, S., Liu, X., Wu, Y., Dong, F., Qiu, C.W. and Qiu, J., 2021. Artificial intelligence: A powerful paradigm for scientific research. *The Innovation*, *2*(4).

Zhang, C. and Lu, Y., 2021. Study on artificial intelligence: The state of the art and future prospects. *Journal of Industrial Information Integration*, *23*, p.100224.

Chapter 6

Supervised and Unsupervised Learning

Supervised vs. Unsupervised Learning

Supervised and Unsupervised learning are the two techniques of machine learning. But both the techniques are used in different scenarios and with different datasets.

Supervised Learning	*Unsupervised Learning*
Supervised learning algorithms are trained using labeled data.	Unsupervised learning algorithms are trained using unlabeled data.
Supervised learning model takes direct feedback to check if it is predicting correct output or not.	Unsupervised learning model does not take any feedback.
Supervised learning model predicts the output.	Unsupervised learning model finds the hidden patterns in data.
In supervised learning, input data is provided to the model along with the output.	In unsupervised learning, only input data is provided to the model.
The goal of supervised learning is to train the model so that it can predict the output when it is given new data.	The goal of unsupervised learning is to find the hidden patterns and useful insights from the unknown dataset.
Supervised learning needs supervision to train the model.	Unsupervised learning does not need any supervision to train the model.
Supervised learning can be categorized in **Classification** and **Regression** problems.	Unsupervised Learning can be classified in **Clustering** and **Associations** problems.

Supervised Learning	*Unsupervised Learning*
Supervised learning can be used for those cases where we know the input as well as corresponding outputs.	Unsupervised learning can be used for those cases where we have only input data and no corresponding output data.
Supervised learning model produces an accurate result.	Unsupervised learning model may give less accurate result as compared to supervised learning.
Supervised learning is not close to true Artificial intelligence as in this, we first train the model for each data, and then only it can predict the correct output.	Unsupervised learning is more close to the true Artificial Intelligence as it learns similarly as a child learns daily routine things by his experiences.
It includes various algorithms such as Linear Regression, Logistic Regression, Support Vector Machine, Multi-class Classification, Decision tree, Bayesian Logic, *etc.*	It includes various algorithms such as Clustering, KNN, and Apriori algorithm.

These two are the broad categories of machine learning model approach

Supervised Learning

Supervised learning, as the name indicates, has the presence of a supervisor as a teacher. Basically supervised learning is when we teach or train the machine using data that is well-labelled. Which means some data is already tagged with the correct answer. After that, the machine is provided with a new set of examples(data) so that the supervised learning algorithm analyses the training data(set of training examples) and produces a correct outcome from labeled data.

Supervised learning is classified into two categories of algorithms:

Classification: A classification problem is when the output variable is a category, such as "Red" or "blue", "disease" or "no disease".

Regression: A regression problem is when the output variable is a real value, such as "dollars" or "weight".

Supervised learning deals with or learns with "labeled" data. This implies that some data is already tagged with the correct answer.

Types

- Regression
- Logistic Regression
- Classification
- Naive Bayes Classifiers
- K-NN (k nearest neighbors)
- Decision Trees
- Support Vector Machine

Unsupervised Learning

Unsupervised learning is the training of a machine using information that is neither classified nor labeled and allowing the algorithm to act on that information without guidance. Here the task of the machine is to group unsorted information according to similarities, patterns, and differences without any prior training of data. Unlike supervised learning, no teacher is provided that means no training will be given to the machine. Therefore, the machine is restricted to find the hidden structure in unlabeled data by itself.

Unsupervised learning is classified into two categories of algorithms:

Clustering: A clustering problem is where you want to discover the inherent groupings in the data, such as grouping customers by purchasing behavior.

Association: An association rule learning problem is where you want to discover rules that describe large portions of your data, such as people that buy X also tend to buy Y.

Types

- Hierarchical clustering
- K-means clustering
- Principal Component Analysis

Below the explanation of both learning methods along with their difference table is given.

Classification

Classification is a supervised machine learning method where the model tries to predict the correct label of a given input data. In classification, the model is fully trained using the training data, and then it is evaluated on test data before being used to perform prediction on new unseen data.

Why do we need to do Classification?

For many applications, it is important to know what type of land cover pixel represents. (*i.e.* water)

For a given land cover, we often need to know about different types of land uses (*i.e.* agriculture vs. forest)

We also need to quantify land use patterns (*i.e.* how much agricultural area in a region)

For detecting changes, we need to compare how a pixel changed from Class A to Class B

The main classification problems include:

1. Binary classification
2. Multiple class classification
3. Multiple label classification

Binary classification: Binary is a type of problem in classification in machine learning that has only two possible outcomes. For example, yes or no, true or false, spam or not spam, *etc.*

Examples include: Logistic Regression, Decision Trees, Naïve Bayes

Multiple class classification: Multi-class is a type of classification problem with more than two outcomes and does not have the concept of normal and abnormal outcomes. Here each outcome is assigned to only one label. For example, classifying images, classifying species, and categorizing faces, among others.

Examples include: decision Tree, Random Forest, k-Nearest Neighbors, Naive Bayes, Artificial neural network(ANN)

Multiple label classification: Multi-label is a type of classification problem that may have more than one class label assigned to the data. Here the model will have multiple outcomes. For example, a book or a movie can be categorized into multiple genres, or an image can have multiple objects.

Examples: Multiple label Gradient Boosting, Multiple label Random Forests, Using different classification algorithms for each class label

The general workflow for classification is:

- Collect training data.
- Assemble features which have a property that stores the known class label and properties storing numeric values for the predictors.
- Instantiate a classifier.
- Set its parameters if necessary.

- Train the classifier using the training data.
- Classify an image or feature collection.
- Estimate classification error with independent validation data.

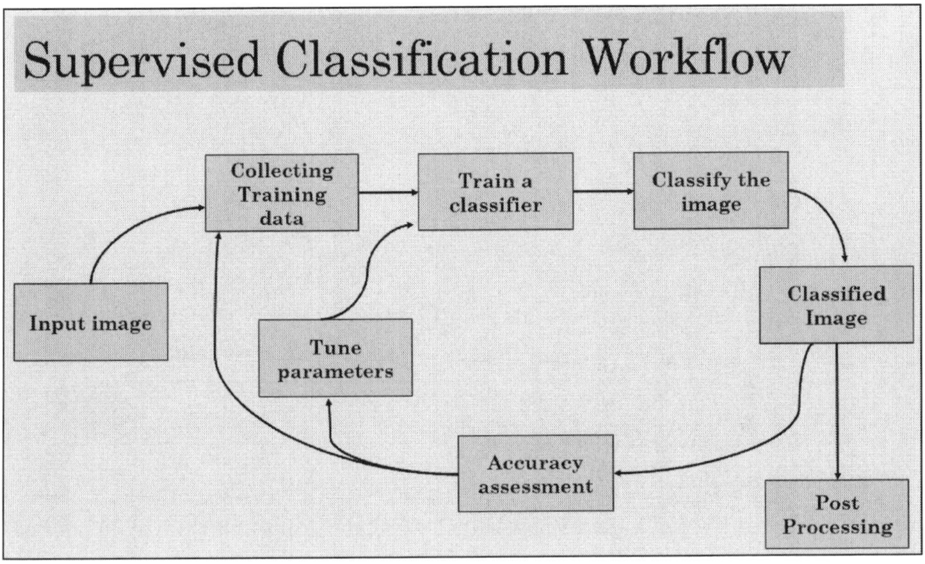

Different Supervised ML Algorithms

- Support vector machine
- Decision Tree
- KNN (K Nearest Neighbour)

Support Vector Machines

Support Vector Machine or SVM is one of the most popular Supervised Learning algorithms, which is used for Classification as well as Regression problems. However, primarily, it is used for Classification problems in Machine Learning. The goal of the SVM algorithm is to create the best line or decision boundary that can segregate n-dimensional space into classes so that we can easily put the new data point in the correct category in the future. This best decision boundary is called a hyperplane. SVM chooses the extreme points/vectors that help in creating the hyperplane. These extreme cases are called as support vectors, and hence algorithm is termed as Support Vector Machine. Consider the below diagram in which there are two different categories that are classified using a decision boundary or hyperplane:

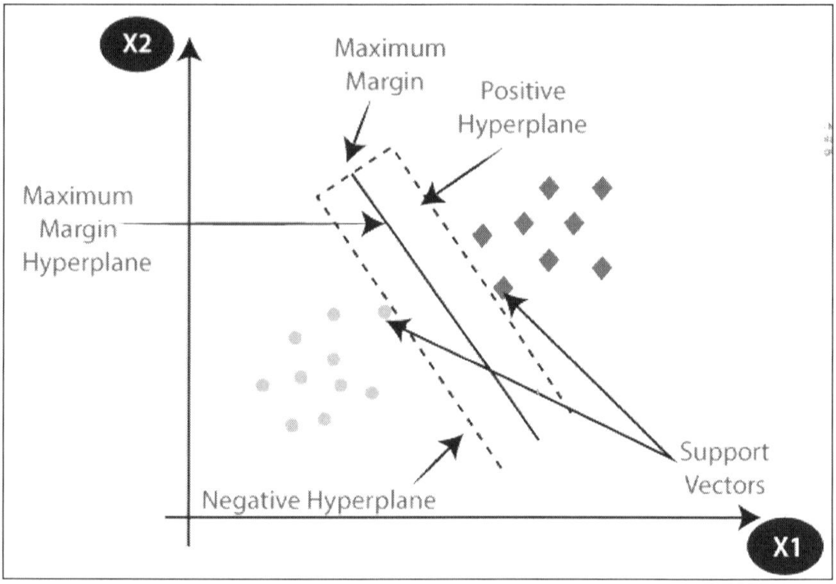

SVM can be of two types:

- **Linear SVM:** Linear SVM is used for linearly separable data, which means if a dataset can be classified into two classes by using a single straight line, then such data is termed as linearly separable data, and classifier is used called as Linear SVM classifier.
- **Non-linear SVM:** Non-Linear SVM is used for non-linearly separated data, which means if a dataset cannot be classified by using a straight line, then such data is termed as non-linear data and classifier used is called as Non-linear SVM classifier.

Decision Trees

Decision trees are classified as supervised learning methods that can solve both classification problems and regression problems; however, they are most used to solve classification problems. A tree-based classifier is a classifier in which each leaf node is the classification result and inside nodes are the features of the dataset.

Following figure represents decision tree consisting of two nodes: the Decision node and the Leaf node. A Decision node is a decision that has several branches. A Leaf node, on the other hand, is a result of this decision and has no additional branches. A decision tree is a graphical representation of all possible answers to a question or problem based on a set of conditions. The name decision tree comes from the fact that its structure is tree-like; it starts at the root node and branches grow on top of each other to form a tree-like structure. The algorithm is used to construct a tree. A decision tree just asks a question and divides a tree into subtrees based on the answer (Yes/No).

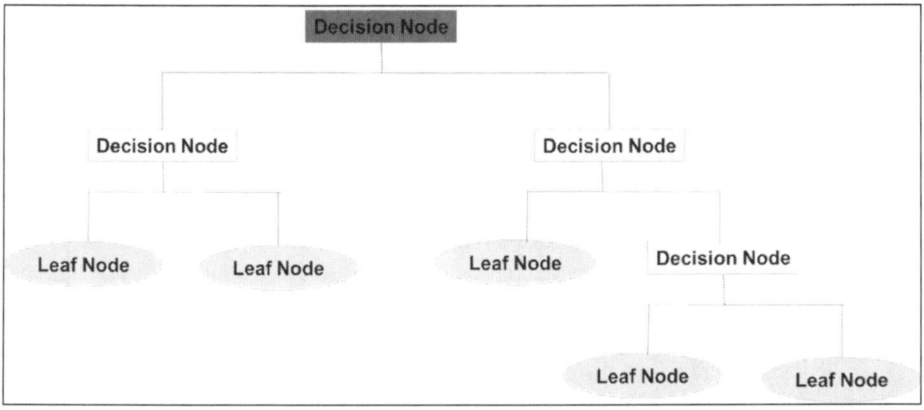

Above Figure represents the basic structure of a decision tree

Working procedure of decision tree in a decision tree, the algorithm begins at the root node and works its way up to forecast the class of the given dataset. This algorithm follows the branch and jumps to the following node by comparing the values of the root attribute with those of the record (real dataset) attribute. The algorithm verifies the attribute value with the other sub-nodes once again for the following node before continuing. This process continues until it reaches the leaf node of the decision tree.

- Step-1: Start with the root node of the tree, which has the entire dataset.
- Step-2: Using Attribute Selection Measure (ASM), identify the top attribute in the dataset.
- Step-3: Divide the root node into subsets that contain possible values for the best attributes.
- Step-4: Create the best attribute-containing decision tree node.
- Step-5: New decision trees will be iteratively designed using the subsets of the dataset obtained in step 3. This process will be continued until the model reaches a point where the nodes can no longer be categorized, and the last node will be referred to as a leaf node.

Types of Decision Trees in Machine Learning

Decision trees in machine learning can either be classification trees or regression trees. Together, both types of algorithms fall into a category of "classification and regression trees" and are sometimes referred to as CART. Their respective roles are to "classify" and to "predict."

1. Classification Trees

Classification trees determine whether an event happened or didn't happen. Usually, this involves a "yes" or "no" outcome.

We often use this type of decision-making in the real world. Here are a few examples to help contextualize how decision trees work for classification:

Example 1: How to spend your free time after work

What you do after work in your free time can be dependent on the weather. If it is sunny, you might choose between having a picnic with a friend, grabbing a drink with a colleague, or running errands. If it is raining, you might opt to stay home and watch a movie instead. There is a clear outcome. In this case, that is classified as whether to "go out" or "stay in."

2. Regression Tree

Regression trees, on the other hand, predict continuous values based on previous data or information sources. For example, they can predict the price of gasoline or whether a customer will purchase eggs (including which type of eggs and at which store).

This type of decision-making is more about programming algorithms to predict what is likely to happen, given previous behavior or trends.

Example 1: Housing prices in Colorado

Regression analysis could be used to predict the price of a house in Colorado, which is plotted on a graph. The regression model can predict housing prices in the coming years using data points of what prices have been in previous years. This relationship is a linear regression since housing prices are expected to continue rising. Machine learning helps us predict specific prices based on a series of variables that have been true in the past.

To get started on how decision tree algorithms work in predictive machine learning models, take a look at these guided projects. Each project takes less than two hours, and they are based on real-world examples so you can elevate your skills:

Classification and Regression Tree (CART) is a predictive algorithm used in machine learning that generates future predictions based on previous values. These decision trees are at the core of machine learning, and serve as a basis for other machine learning algorithms such as random forest, bagged decision trees, and boosted decision trees.

Advantages

- It produces a highly accurate classifier and learning is fast.
- It runs efficiently on large data bases.
- It can handle thousands of input variables without variable deletion.
- It computes proximities between pars of cases that can be used in clustering, locating outliers or give interesting views of the data.
- It offers an experimental method for detecting variable interactions.

Random Forest

Random Forest is a popular machine learning algorithm that belongs to the supervised learning technique. It can be used for both Classification and Regression problems in ML. It is based on the concept of ensemble learning, which is a process of combining multiple classifiers to solve a complex problem and to improve the performance of the model.

As the name suggests, "Random Forest is a classifier that contains a number of decision trees on various subsets of the given dataset and takes the average to improve the predictive accuracy of that dataset." Instead of relying on one decision tree, the random forest takes the prediction from each tree and based on the majority votes of predictions, and it predicts the final output.

The greater number of trees in the forest leads to higher accuracy and prevents the problem of overfitting.

The below diagram explains the working of the Random Forest algorithm:

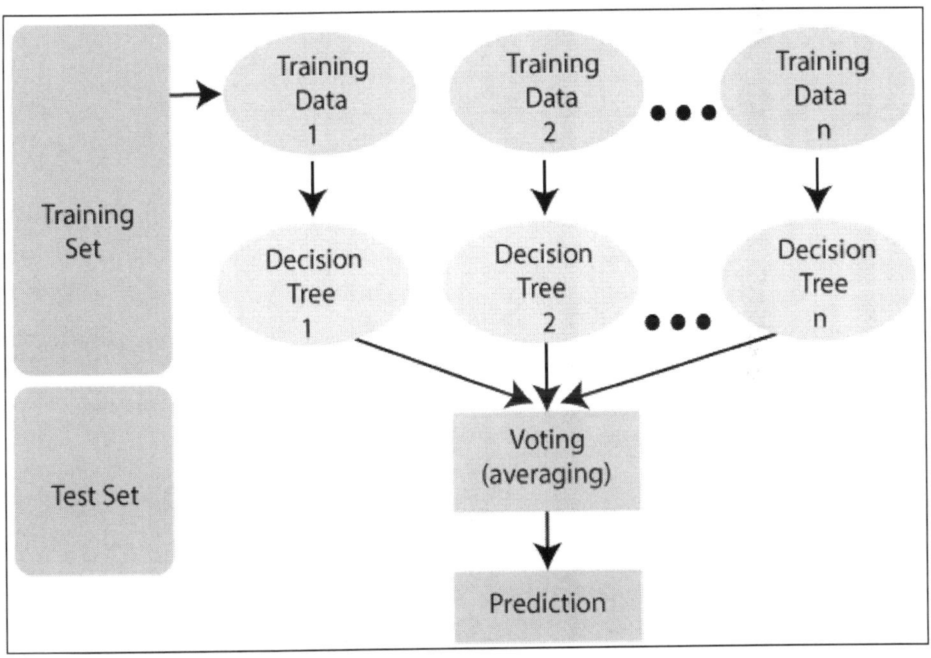

How does Random Forest algorithm work?

Random Forest works in two-phase first is to create the random forest by combining N decision tree, and second is to make predictions for each tree created in the first phase.

The Working process can be explained in the below steps and diagram:

Step-1: Select random K data points from the training set.

Step-2: Build the decision trees associated with the selected data points (Subsets).

Step-3: Choose the number N for decision trees that you want to build.

Step-4: Repeat Step 1 and 2.

Step-5: For new data points, find the predictions of each decision tree, and assign the new data points to the category that wins the majority votes.

The working of the algorithm can be better understood by the below example:

Example: Suppose there is a dataset that contains multiple fruit images. So, this dataset is given to the Random forest classifier. The dataset is divided into subsets and given to each decision tree. During the training phase, each decision tree produces a prediction result, and when a new data point occurs, then based on the majority of results, the Random Forest classifier predicts the final decision. Consider the below image:

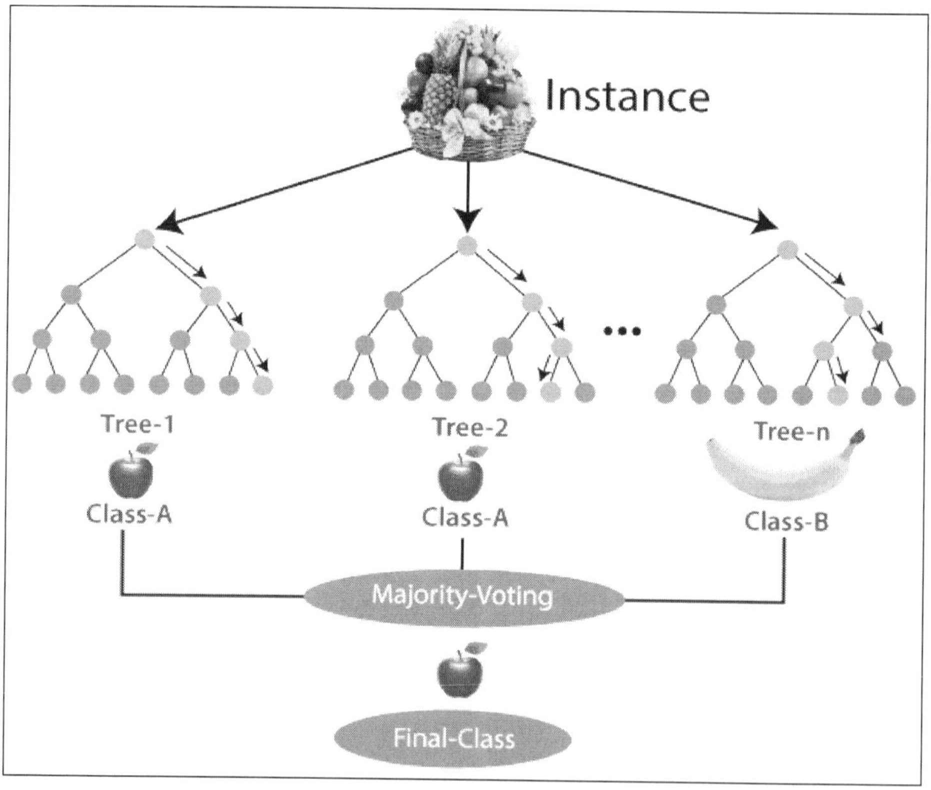

Applications of Random Forest

There are mainly four sectors where Random forest mostly used:

☆ Banking: Banking sector mostly uses this algorithm for the identification of loan risk.

- Medicine: With the help of this algorithm, disease trends and risks of the disease can be identified.
- Land Use: We can identify the areas of similar land use by this algorithm.
- Marketing: Marketing trends can be identified using this algorithm.
- Stock Market: Financial analysts use it to identify potential markets for stocks. It also enables them to remember the behaviour of stocks.
- E-Commerce: Through this system, e-commerce vendors can predict the preference of customers based on past consumption behaviour.

K Nearest Neighbour (KNN)

Fix and Hodges introduced a non-parametric method for pattern classification that has since become known as the k-nearest neighbour rule (Fix and Hodges, 1951). The k-nearest neighbour algorithm is a non-parametric, supervised learning classifier. It uses proximity to rank or predict how a collection of individual points will be sorted, while it can be used for classification or regression problems, it is mostly used as a classifier because it is based on the notion that similar points can be found near each other. For classification problems, a class label is selected by a majority vote. This means that the most expressed label around a given data point is selected. Although this is officially "pluralality voting", literature more commonly refers to it as "majority vote". Like classifications, regression issues make use of the concept; but in this case the average of k nearest neighbours predicts a classification. The main difference is that in discrete data, classification is used,

Whereas in continuous values, regression is used. Defining the distance is required before determining the classification. Euclidean distances are most used. Figure represents a basic example of KNN model.

Euclidean Distance (P=2)

This is the commonly used method to find the distance between two points in KNN. And this approach can measure only real-valued vectors. P in the Euclidean distance formula is the order of the norm

The formula is:

$$Euclidean\ Distance, d(x,y) = \sqrt{\sum_{i=1}^{n}(y_i - x_i)^2}$$

Manhattan Distance (p=1)

The absolute value between two places is measured using this distance metric.

$$Manhattan\ distance, d(x,y) = \sum_{i=1}^{n} |x_i - y_i|$$

Minkowski Distance

It is the generalized version of the Euclidean and Manhattan distance metrics.

$$Minkowski\ distance, d(x,y) = \sum_{i=1}^{n} |x_i - y_i|^{1/p}$$

Hamming Distance

This method is frequently used to locate the spots where two vectors do not match, whether they be a string or Boolean vectors.

$$Hamming\ distance\ (D_H) = \sum_{i=1}^{k} |x_i - y_i|$$

- ☆ Step-1: Choose the neighbour with the K numbers.
- ☆ Step-2: Determine the Euclidean distance between K neighbours.
- ☆ Step-3: Take the K nearest neighbours as per the calculated Euclidean distance.
- ☆ Step-4: Count how many data points there are in each category among these k neighbours.
- ☆ Step-5: Assign the newly added data points to the category where the neighbour count is at its highest.
- ☆ Step-6: The model is ready.

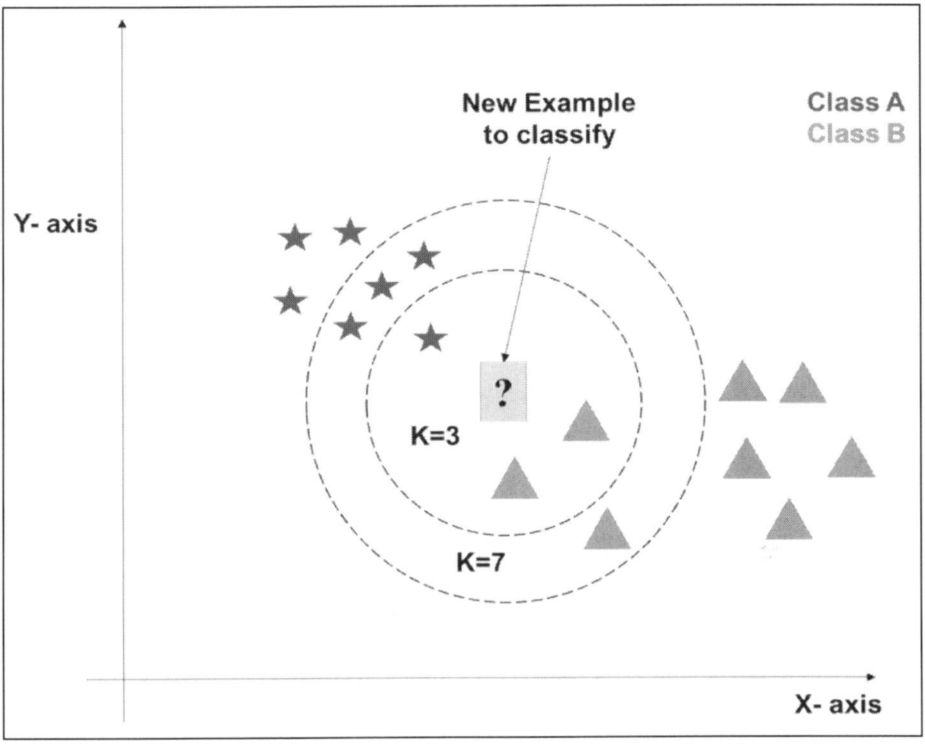

K-nearest Neighbour classification example for K=3 and K=7

Example: Suppose, we have an image of a creature that looks similar to cat and dog, but we want to know either it is a cat or dog. So for this identification, we can use the KNN algorithm, as it works on a similarity measure. Our KNN model will find the similar features of the new data set to the cats and dogs images and based on the most similar features it will put it in either cat or dog category.

Advantages of KNN Algorithm

- It is simple to implement.
- It is robust to the noisy training data
- It can be more effective if the training data is large.

Naïve Bayes Classifier

A simplistic probability classifier, the Naive Bayes method determines a set of probabilities by counting the frequency and combinations of values in each data set. Calculating conditional probability is based on the Bayes theorem, named after Thomas Bayes the British mathematician. The Bayes algorithm takes the class variable as its parameter and supposes that all the variables are independent of each other. The Bayes method learns quickly in many controlled classification problems. However, this conditional independence is a weak assumption as it rarely applies in real-life situations. (Dimitoglou *et al.,* 2012)

$$P(A|B) = \frac{P(A)P(B|A)}{P(B)}$$

Bayes classifiers are simple probabilistic models based on Bayes' theorem and weak (no-obligation) independence assumptions. The phrase "independent feature model" might be a more apt description of the underlying probabilistic model.

A Bayesian classifier is called a "naive Bayes model" because it assumes that the presence or absence of one character within a class does not affect the presence or absence (or not) of another feature. For example, a red apple is classified as an apple if it is rounded, red, and has a diameter of about 4". A naive Bayesian classifier supposes that each of those characteristics independently contributes to the probability that this apple is an apple (even if they all interact with each other or the presence of other features).

A Bayesian classifier may be trained relatively well in your supervised learning environment, depending on the specifics of your probability model. It is possible to work with the Bayesian model without Bayesian probability and no Bayesian techniques, as parameter estimation for Bayesian models is often done using a maximum likelihood method. The corresponding classifier is the function classify defined as follows:

$$Classify(f_1, \dots \dots, f_n) = \underset{c}{argmax}\ p(C = c)\prod_{i=1}^{n} p(F_i = f_1|C = c)$$

This means that this needs to add the conditional probability for each feature for each class label. The individual conditional probability for each class label and for each feature is calculated, and it is multiplied by the prior conditional probability for that class label (P(Fi)|Cj). The label that is returned by the classifier is the label for which the best result is obtained

Logistic Regression

Logistic regression is a supervised machine learning algorithm used for classification tasks where the goal is to predict the probability that an instance belongs to a given class or not. Logistic regression is a statistical algorithm which analyze the relationship between two data factors. The article explores the fundamentals of logistic regression, it's types and implementations.

What is Logistic Regression?

Logistic regression is used for binary classification where we use sigmoid function, that takes input as independent variables and produces a probability value between 0 and 1.

For example, we have two classes Class 0 and Class 1 if the value of the logistic function for an input is greater than 0.5 (threshold value) then it belongs to Class 1 otherwise it belongs to Class 0. It's referred to as regression because it is the extension of linear regression but is mainly used for classification problems.

Key Points

- Logistic regression predicts the output of a categorical dependent variable. Therefore, the outcome must be a categorical or discrete value.
- It can be either Yes or No, 0 or 1, true or False, *etc.* but instead of giving the exact value as 0 and 1, it gives the probabilistic values which lie between 0 and 1.
- In Logistic regression, instead of fitting a regression line, we fit an "S" shaped logistic function, which predicts two maximum values (0 or 1).

Logistic Function – Sigmoid Function

The sigmoid function is a mathematical function used to map the predicted values to probabilities. It maps any real value into another value within a range of 0 and 1. The value of the logistic regression must be between 0 and 1, which cannot go beyond this limit, so it forms a curve like the "S" form.

The S-form curve is called the Sigmoid function or the logistic function.

In logistic regression, we use the concept of the threshold value, which defines the probability of either 0 or 1. Such as values above the threshold value tends to 1, and a value below the threshold values tends to 0.

Types of Logistic Regression

On the basis of the categories, Logistic Regression can be classified into three types:

- Binomial: In binomial Logistic regression, there can be only two possible types of the dependent variables, such as 0 or 1, Pass or Fail, *etc.*
- Multinomial: In multinomial Logistic regression, there can be 3 or more possible unordered types of the dependent variable, such as "cat", "dogs", or "sheep"
- Ordinal: In ordinal Logistic regression, there can be 3 or more possible ordered types of dependent variables, such as "low", "Medium", or "High".

Assumptions of Logistic Regression

We will explore the assumptions of logistic regression as understanding these assumptions is important to ensure that we are using appropriate application of the model. The assumption include:

- Independent observations: Each observation is independent of the other. Meaning there is no correlation between any input variables.
- Binary dependent variables: It takes the assumption that the dependent variable must be binary or dichotomous, meaning it can take only two values. For more than two categories SoftMax functions are used.
- Linearity relationship between independent variables and log odds: The relationship between the independent variables and the log odds of the dependent variable should be linear.
- No outliers: There should be no outliers in the dataset.
- Large sample size: The sample size is sufficiently large
- Terminologies involved in Logistic Regression

Here are some common terms involved in logistic regression:

- Independent variables: The input characteristics or predictor factors applied to the dependent variable's predictions.
- Dependent variable: The target variable in a logistic regression model, which we are trying to predict.
- Logistic function: The formula used to represent how the independent and dependent variables relate to one another. The logistic function transforms the input variables into a probability value between 0 and 1, which represents the likelihood of the dependent variable being 1 or 0.
- Odds: It is the ratio of something occurring to something not occurring. it is different from probability as the probability is the ratio of something occurring to everything that could possibly occur.
- Log-odds: The log-odds, also known as the logit function, is the natural logarithm of the odds. In logistic regression, the log odds of the dependent variable are modeled as a linear combination of the independent variables and the intercept.

- Coefficient: The logistic regression model's estimated parameters, show how the independent and dependent variables relate to one another.
- Intercept: A constant term in the logistic regression model, which represents the log odds when all independent variables are equal to zero.
- Maximum likelihood estimation: The method used to estimate the coefficients of the logistic regression model, which maximizes the likelihood of observing the data given the model.

Logistic Regression Equation

Here x_i is the ith observation of X, $w_i = [w_1, w_2, w_3$,ïe,wm] is the weights or Coefficient, a and b is the bias term also known as intercept. simply this can be represented as the dot product of weight and bias.

z = wX+b

The odd is the ratio of something occurring to something not occurring. it is different from probability as the probability is the ratio of something occurring to everything that could possibly occur so odd will be:

$$\frac{p(x)}{1-p(x)} = e^z$$

Applying natural log on odd. then log odd will be:

$$\log\left[\frac{p(x)}{1-p(x)}\right] = z$$

$$\log\left[\frac{p(x)}{1-p(x)}\right] = w \cdot X + b$$

$$\frac{p(x)}{1-p(x)} = e^{w \cdot X + b} \quad \cdots \text{Exponentiate both sides}$$

$$p(x) = e^{w \cdot X + b} \cdot (1 - p(x))$$

$$p(x) = e^{w \cdot X + b} - e^{w \cdot X + b} \cdot p(x))$$

$$p(x) + e^{w \cdot X + b} \cdot p(x)) = e^{w \cdot X + b}$$

$$p(x)(1 + e^{w \cdot X + b}) = e^{w \cdot X + b}$$

$$p(x) = \frac{e^{w \cdot X + b}}{1 + e^{w \cdot X + b}}$$

then the final logistic regression equation will be:

$$p(X; b, w) = \frac{e^{w \cdot X + b}}{1 + e^{w \cdot X + b}} = \frac{1}{1 + e^{-(w \cdot X + b)}}$$

How to Evaluate Logistic Regression Model?

We can evaluate the logistic regression model using the following metrics:

- **Accuracy:** Accuracy provides the proportion of correctly classified instances.

  Accuracy=TruePositives+TrueNegatives/(total)

- **Precision:** Precision focuses on the accuracy of positive predictions.

  Precision=TruePositives/(TruePositives+FalsePositives)

- **Recall (Sensitivity or True Positive Rate):** Recall measures the proportion of correctly predicted positive instances among all actual positive instances.

  Recall=TruePositives/(TruePositives+FalseNegatives)

- **F1 Score:** F1 score is the harmonic mean of precision and recall.

  F1 Score=2 Precision*Recall/(Precision+Recall)

Multinomial Logistic Regression

Multinomial Logistic Regression is similar to logistic regression but with a difference, that the target dependent variable can have more than two classes *i.e.* multiclass or polychotomous.

For example, the students can choose a major for graduation among the streams "Science", "Arts" and "Commerce", which is a multiclass dependent variable and the independent variables can be marks, grade in competitive exams, Parents profile, interest *etc.*

What is Multinomial Logistic Regression?

Multinomial Logistic Regression is a classification technique that extends the logistic regression algorithm to solve multiclass possible outcome problems, given one or more independent variables.

This model is used to predict the probabilities of categorically dependent variable, which has two or more possible outcome classes. Whereas the logistic regression model is used when the dependent categorical variable has two outcome classes for example, students can either "Pass" or "Fail" in an exam or bank manager can either "Grant" or "Reject" the loan for a person. Check out the logistic regression algorithm course and understand this topic in depth.

Example for Multinomial Logistic Regression

Which Flavor of ice cream will a person choose?

- Dependent Variable:

  Vanilla

Chocolate

Butterscotch

Black Current

- ✰ Independent Variables:

  Gender

  Age

  Occasion

  Happiness

  etc.

Multinomial Logistic Regression is also known as multiclass logistic regression, softmax regression, polytomous logistic regression, multinomial logit, maximum entropy (MaxEnt) classifier and conditional maximum entropy model.

Dependent Variable

The dependent Variable can have two or more possible outcomes/classes.

The dependent variables are nominal in nature means there is no any kind of ordering in target dependent classes *i.e.* these classes cannot be meaningfully ordered.

The dependent variable to be predicted belongs to a limited set of items defined.

Assumptions

When you want to choose multinomial logistic regression as the classification algorithm for your problem, then you need to make sure that the data should satisfy some of the assumptions required for multinomial logistic regression.

The Dependent variable should be either nominal or ordinal variable.

Nominal variable is a variable that has two or more categories but it does not have any meaningful ordering in them. For example, (a) 3 types of cuisine *i.e.* Indian, Continental and Italian. (b) 5 categories of transport *i.e.* Bus, Car, Train, Ship and Airplane.

Ordinal variable are variables that also can have two or more categories but they can be ordered or ranked among themselves.

For example, Grades in an exam *i.e.* A-excellent, B-Good, C-Needs Improvement and D-Fail. When ordinal dependent variable is present, one can think of ordinal logistic regression.

Set of one or more Independent variables can be continuous, ordinal or nominal.

Continuous variables are numeric variables that can have infinite number of values within the specified range values. For example, age of a person, number of hours students study, income of an person.

References

Berry, M.W., Mohamed, A. and Yap, B.W. eds., 2019. Supervised and unsupervised learning for data science. Springer Nature.

Love, B.C., 2002. Comparing supervised and unsupervised category learning. Psychonomic bulletin and review, 9(4), pp.829-835.

Rajoub, B., 2020. Supervised and unsupervised learning. In Biomedical signal processing and artificial intelligence in healthcare (pp. 51-89). Academic Press.

Garg, V. and Kalai, A.T., 2018. Supervising unsupervised learning. Advances in Neural Information Processing Systems, 31.

Chapter 7

Artificial Neural Network (ANN)

Introduction to Artificial Neural Networks

Artificial Neural Networks are the most popular machine learning algorithms today. The invention of these Neural Networks took place in the 1970s but they have achieved huge popularity due to the recent increase in computation power because of which they are now virtually everywhere. In every application that you use, Neural Networks power the intelligent interface that keeps you engaged.

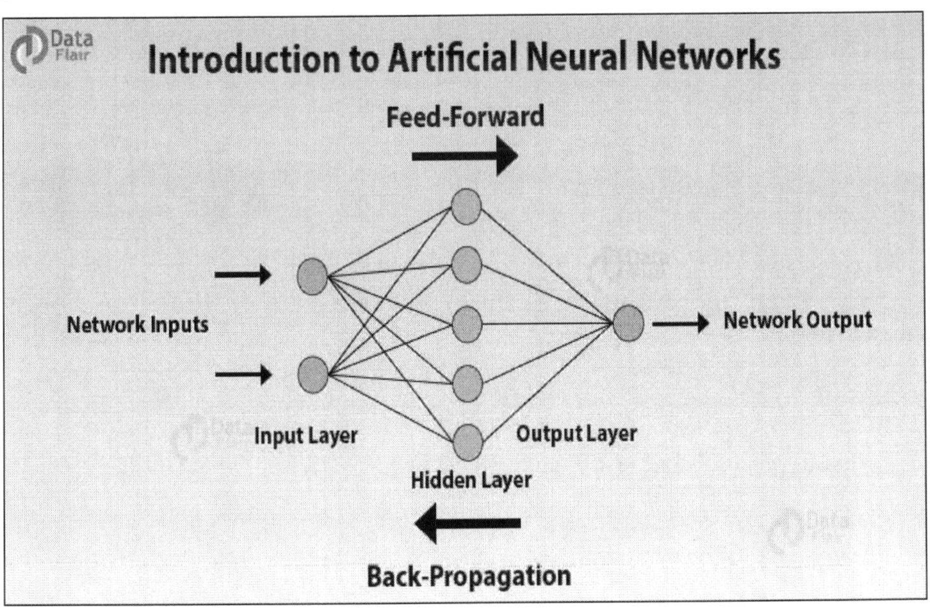

What is ANN?

Artificial Neural Networks are a special type of machine learning algorithms that are modeled after the human brain. That is, just like how the neurons in our nervous system are able to learn from the past data, similarly, the ANN is able to learn from the data and provide responses in the form of predictions or classifications.

ANNs are nonlinear statistical models which display a complex relationship between the inputs and outputs to discover a new pattern. A variety of tasks such as image recognition, speech recognition, machine translation as well as medical diagnosis makes use of these artificial neural networks.

The term "**Artificial Neural Network**" is derived from Biological neural networks that develop the structure of a human brain. Similar to the human brain that has neurons interconnected to one another, artificial neural networks also have neurons that are interconnected to one another in various layers of the networks. These neurons are known as nodes.

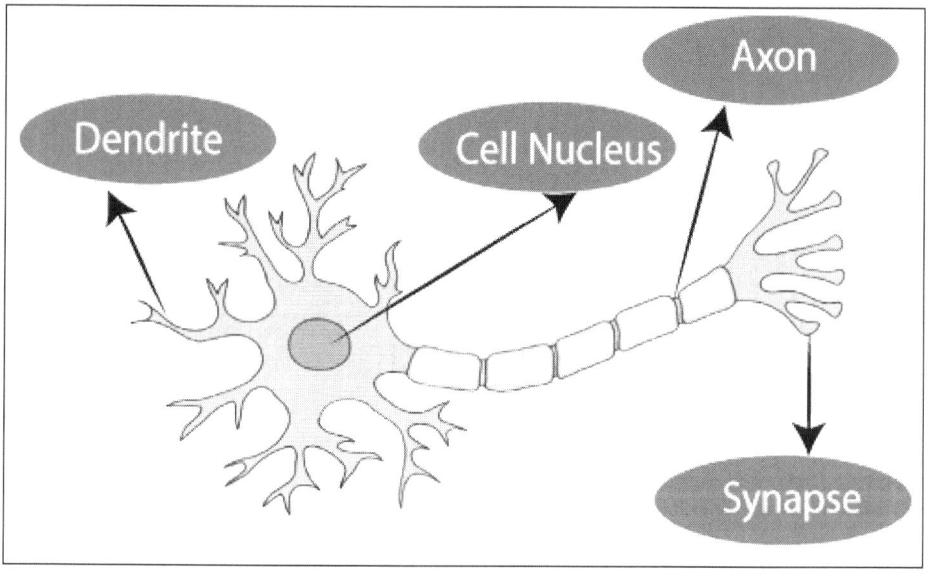

The given figure illustrates the typical diagram of Biological Neural Network.

The typical Artificial Neural Network looks something like the given figure.

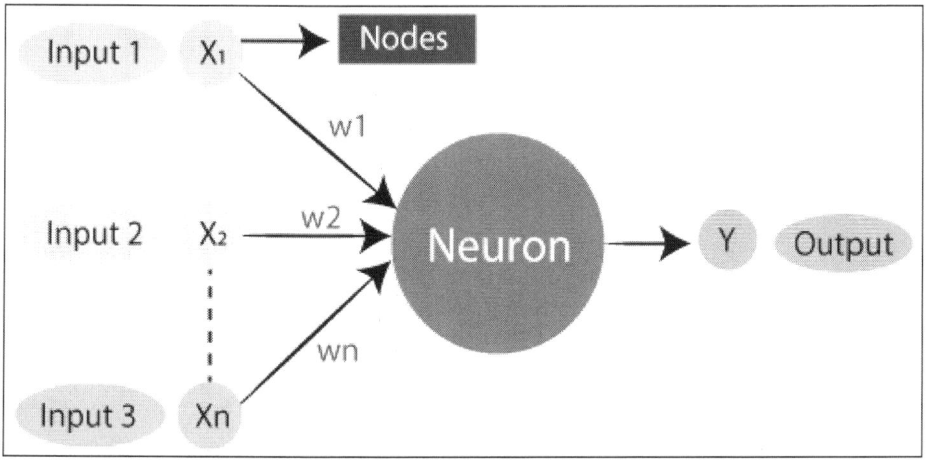

Dendrites from Biological Neural Network represent inputs in Artificial Neural Networks, cell nucleus represents Nodes, synapse represents Weights, and Axon represents Output.

Relationship between Biological neural network and artificial neural network:

Biological Neural Network	*Artificial Neural Network*
Dendrites	Inputs
Cell nucleus	Nodes
Synapse	Weights
Axon	Output

An **Artificial Neural Network** in the field of **Artificial intelligence** where it attempts to mimic the network of neurons makes up a human brain so that computers will have an option to understand things and make decisions in a human-like manner. The artificial neural network is designed by programming computers to behave simply like interconnected brain cells.

An important advantage of ANN is the fact that it learns from the example data sets. Most commonly usage of ANN is that of a random function approximation. With these types of tools, one can have a cost-effective method of arriving at the solutions that define the distribution. ANN is also capable of taking sample data rather than the entire dataset to provide the output result. With ANNs, one can enhance existing data analysis techniques owing to their advanced predictive capabilities.

Artificial Neural Networks Architecture

The functioning of the Artificial Neural Networks is similar to the way neurons work in our nervous system. The Neural Networks go back to the early 1970s when Warren S McCulloch and Walter Pitts coined this term. In order to understand the workings of ANNs, let us first understand how it is structured. In a neural network, there are three essential layers –

- Input Layers: The input layer is the first layer of an ANN that receives the input information in the form of various texts, numbers, audio files, image pixels, *etc.*
- Hidden Layers: In the middle of the ANN model are the hidden layers. There can be a single hidden layer, as in the case of a perceptron or multiple hidden layers. These hidden layers perform various types of mathematical computation on the input data and recognize the patterns that are part of.
- Output Layer: In the output layer, we obtain the result that we obtain through rigorous computations performed by the middle layer.

In a neural network, there are multiple parameters and hyperparameters that affect the performance of the model. The output of ANNs is mostly dependent on these parameters. Some of these parameters are weights, biases, learning rate, batch size *etc.* Each node in the ANN has some weight.

Each node in the network has some weights assigned to it. A transfer function is used for calculating the weighted sum of the inputs and the bias.

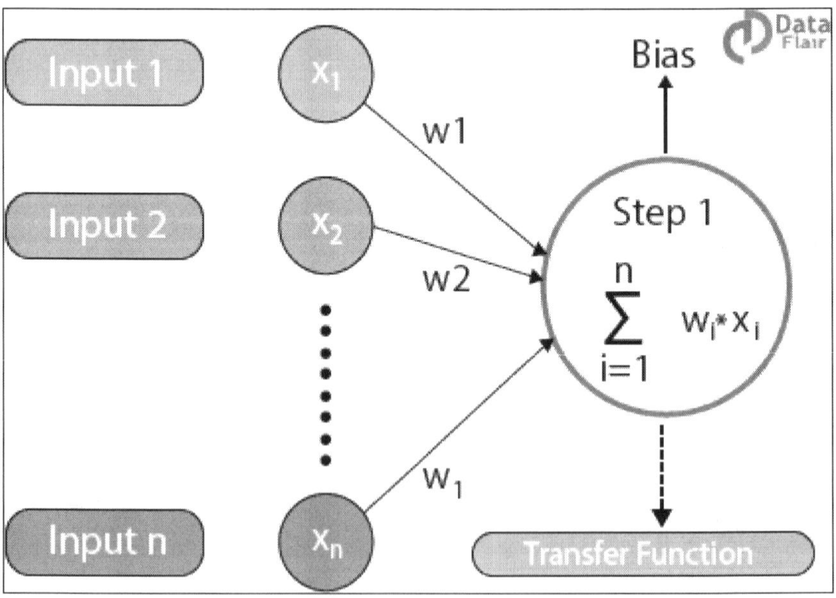

After the transfer function has calculated the sum, the activation function obtains the result. Based on the output received, the activation functions fire the appropriate result from the node. For example, if the output received is above 0.5, the activation function fires a 1 otherwise it remains 0.

Some of the popular activation functions used in Artificial Neural Networks are Sigmoid, RELU, Softmax, tanh *etc.*

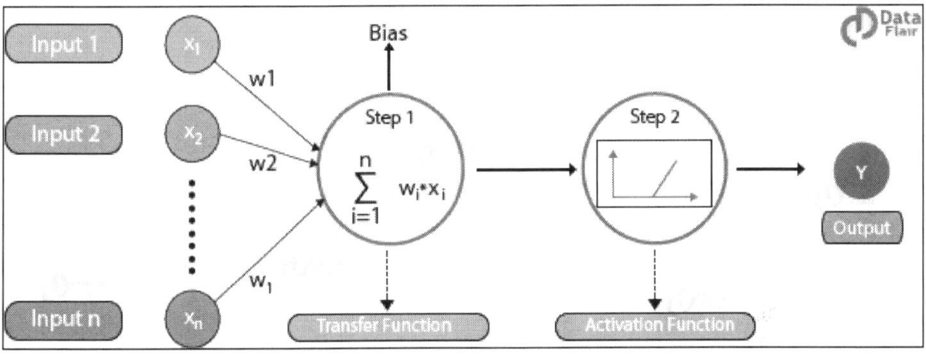

Based on the value that the node has fired, we obtain the final output. Then, using the error functions, we calculate the discrepancies between the predicted output and resulting output and adjust the weights of the neural network through a process known as backpropagation.

ANNs are part of an emerging area in Machine Learning known as Deep Learning.

Many people are confused between Deep Learning and Machine Learning. Are you among one of them? Check this easy to understand article on Deep Learning vs Machine Learning.

Back Propagation in Artificial Neural Networks

In order to train a neural network, we provide it with examples of input-output mappings. Finally, when the neural network completes the training, we test the neural network where we do not provide it with these mappings. The neural network predicts the output and we evaluate how correct the output is using the various error functions. Finally, based on the result, the model adjusts the weights of the neural networks to optimize the network following gradient descent through the chain rule.

Types of Artificial Neural Networks

There are two important types of Artificial Neural Networks –

- FeedForward Neural Network
- FeedBack Neural Network
- FeedForward Artificial Neural Networks

In the feedforward ANNs, the flow of information takes place only in one direction. That is, the flow of information is from the input layer to the hidden layer and finally to the output. There are no feedback loops present in this neural network. These type of neural networks are mostly used in supervised learning for instances such as classification, image recognition *etc.* We use them in cases where the data is not sequential in nature.

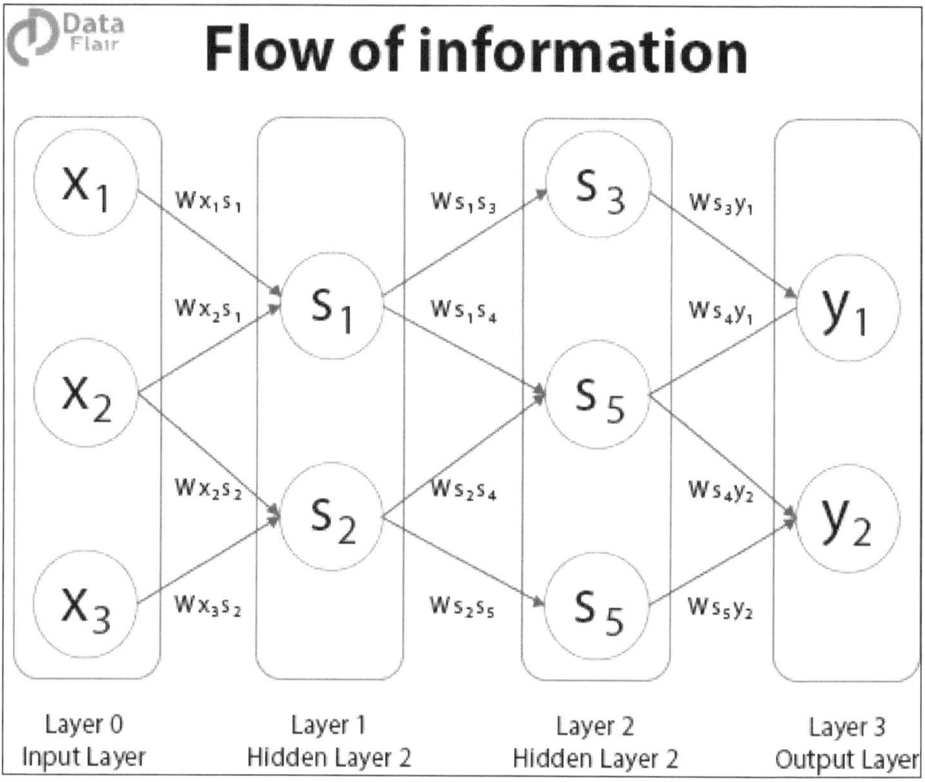

Feedback Artificial Neural Networks

In the feedback ANNs, the feedback loops are a part of it. Such type of neural networks are mainly for memory retention such as in the case of recurrent neural networks. These types of networks are most suited for areas where the data is sequential or time-dependent.

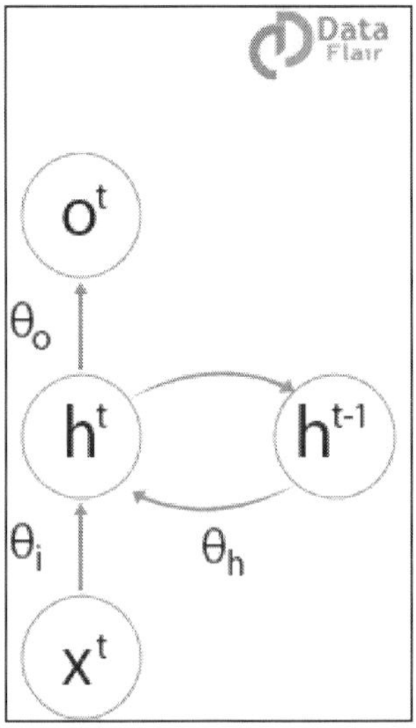

Artificial Neural Networks Applications

Following are the important Artificial Neural Networks applications –

- Handwritten Character Recognition: ANNs are used for handwritten character recognition. Neural Networks are trained to recognize the handwritten characters which can be in the form of letters or digits.

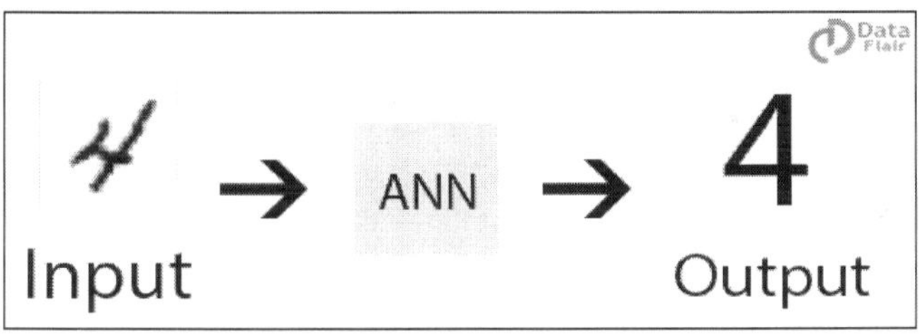

- Speech Recognition: ANNs play an important role in speech recognition. The earlier models of Speech Recognition were based on statistical models like Hidden Markov Models. With the advent of deep learning, various types of neural networks are the absolute choice for obtaining an accurate classification.

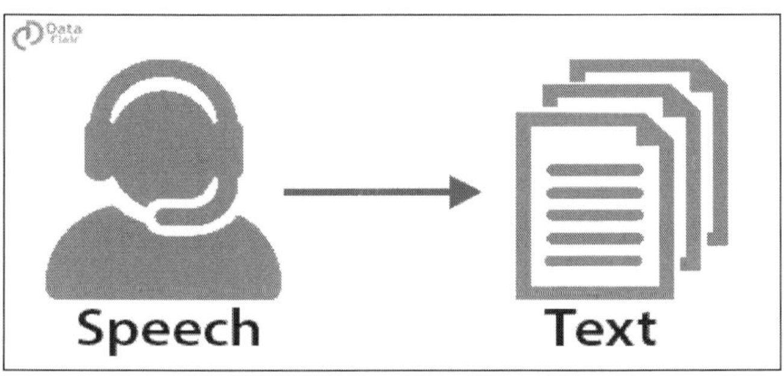

Advantages

- ☆ Parallel processing capability
- ☆ Storing data on the entire network
- ☆ Capability to work with incomplete knowledge
- ☆ Having a memory distribution
- ☆ Having fault tolerance

Chapter 8

Convolutional Neural Network

Introduction to Convolutional Neural Network (CNN)

Convolutional Neural Network is one of the technique to do image classification and image recognition in neural networks. It is designed to process the data by multiple layers of arrays. This type of neural network is used in applications like image recognition or face recognition. The primary difference between CNN and other neural network is that CNN takes input as a two-dimensional array. And it operates directly on the images rather than focusing on feature extraction which other neural networks do.

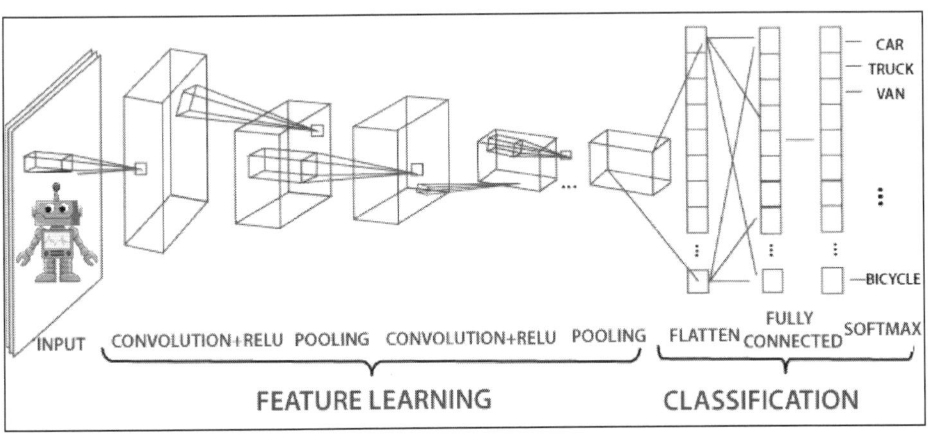

The dominant approach of CNN includes the solution for problems of recognition. Some companies like Google and Facebook have invested in the field research and development concerning recognition projects to get activities done with higher speed.

Scene labeling, object detection, and face recognition, *etc.* are some of the areas where Convolutional Neural Network works.

Convolutional Neural Network (CNN or ConvNet) is a type of feed-forward artificial network where the connectivity pattern between its neurons is inspired by the organization of the animal visual cortex.

Working of CNN

Generally, A Convolutional neural network has three layers. And we understand each layer one by one with the help of an example of the classifier. With it can classify an image of an X and O. So, with the case, we will understand all four layers.

Convolutional Neural Networks have the following layers:

- Convolutional
- ReLU Layer
- Pooling
- Fully connected layer

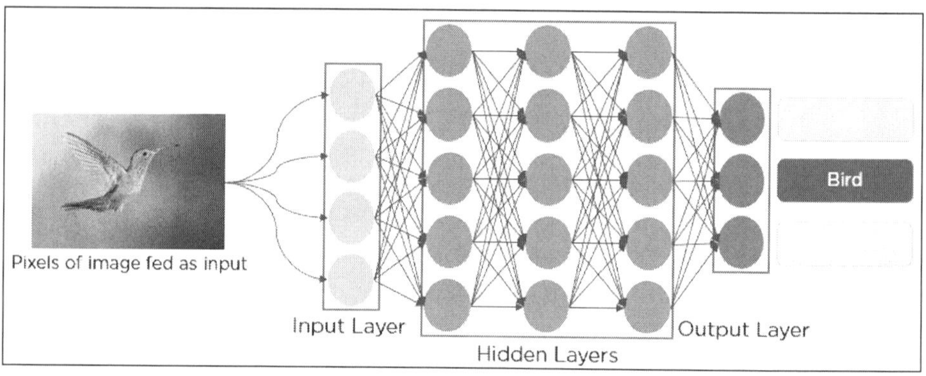

Convolution Layer

This is the first step in the process of extracting valuable features from an image. A convolution layer has several filters that perform the convolution operation. Every image is considered as a matrix of pixel values.

Consider the following 5x5 image whose pixel values are either 0 or 1. There's also a filter matrix with a dimension of 3x3. Slide the filter matrix over the image and compute the dot product to get the convolved feature matrix.

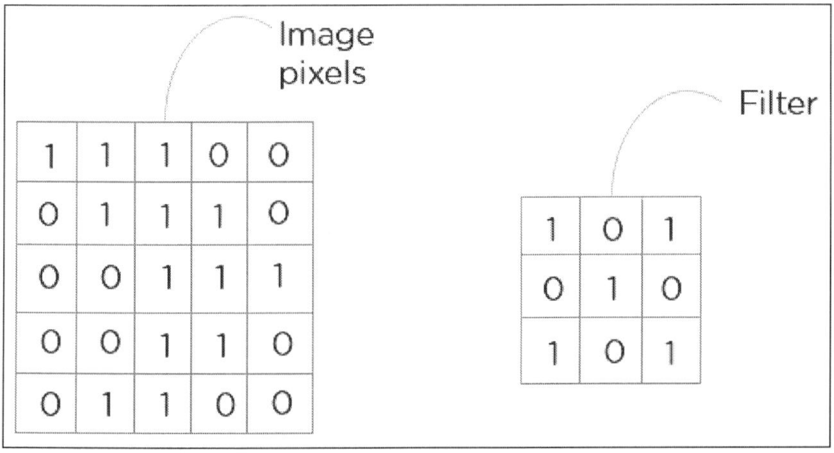

ReLU Layer

ReLU stands for the rectified linear unit. Once the feature maps are extracted, the next step is to move them to a ReLU layer.

ReLU performs an element-wise operation and sets all the negative pixels to 0. It introduces non-linearity to the network, and the generated output is a rectified feature map. Below is the graph of a ReLU function:

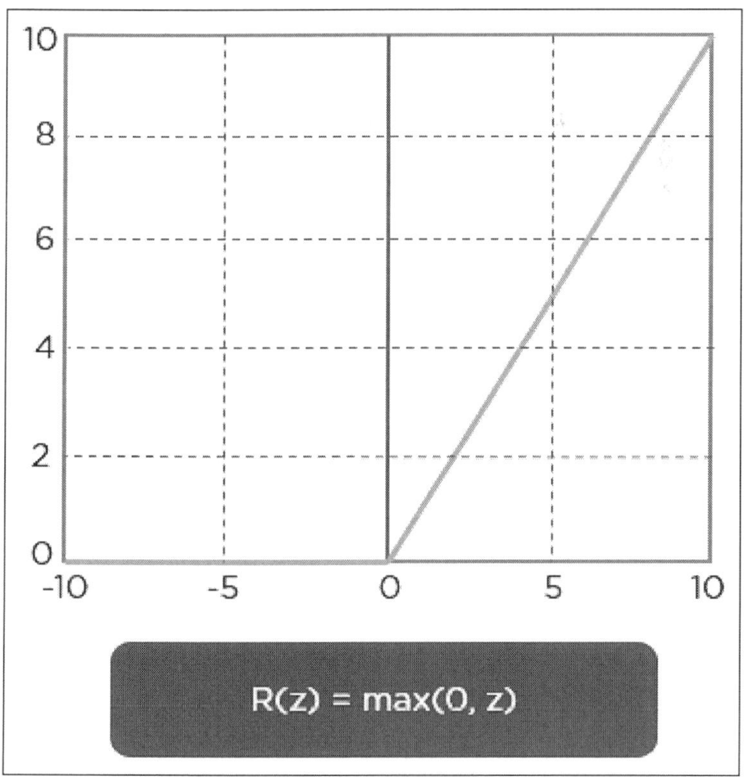

The original image is scanned with multiple convolutions and ReLU layers for locating the features.

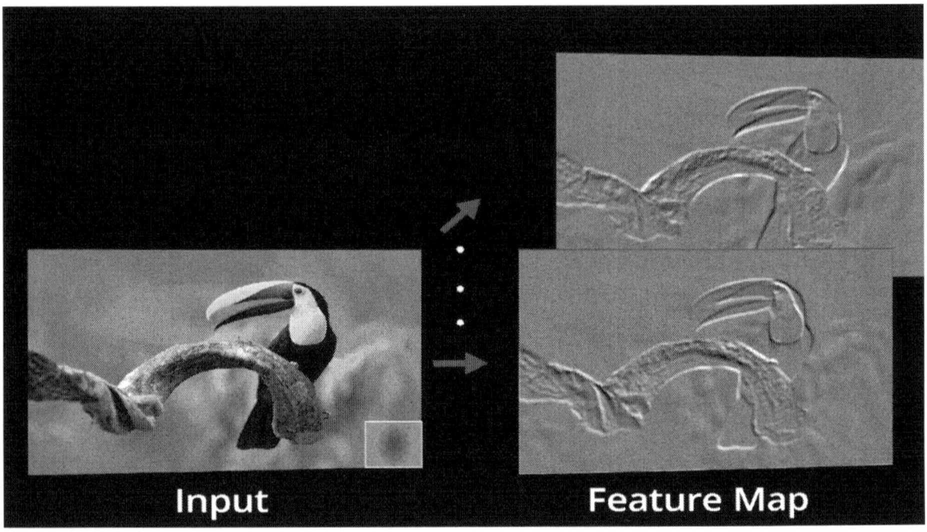

Pooling Layer

Pooling is a down-sampling operation that reduces the dimensionality of the feature map. The rectified feature map now goes through a pooling layer to generate a pooled feature map.

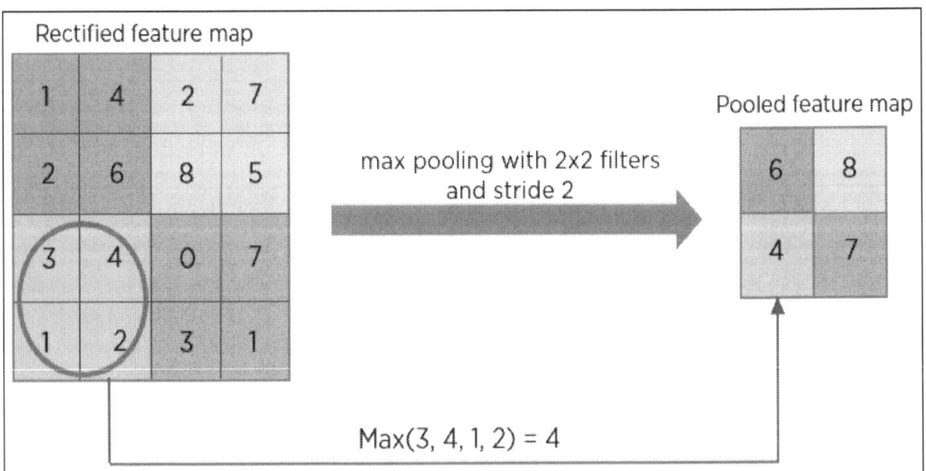

The pooling layer uses various filters to identify different parts of the image like edges, corners, body, feathers, eyes, and beak.

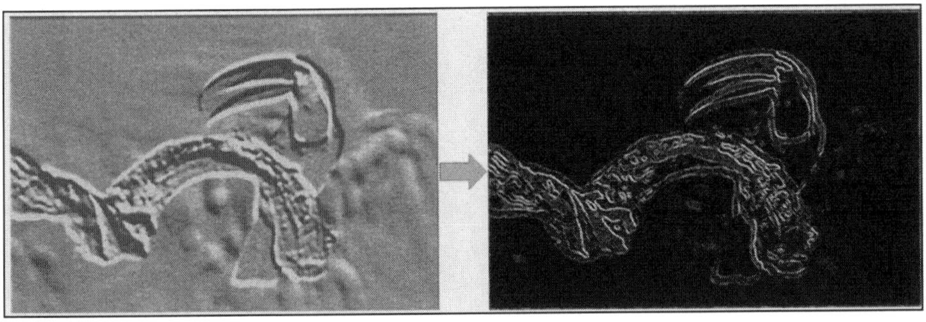

Here's how the structure of the convolution neural network looks so far:

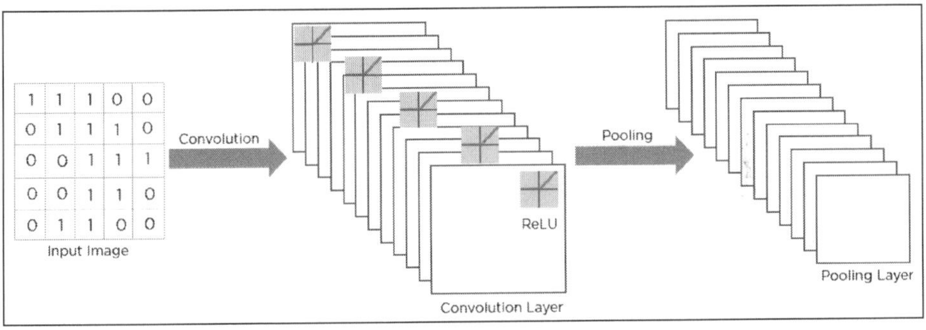

The next step in the process is called flattening. Flattening is used to convert all the resultant 2-Dimensional arrays from pooled feature maps into a single long continuous linear vector.

Fully Connected Layer

The next step in the process is called flattening. Flattening is used to convert all the resultant 2-Dimensional arrays from pooled feature maps into a single long continuous linear vector.

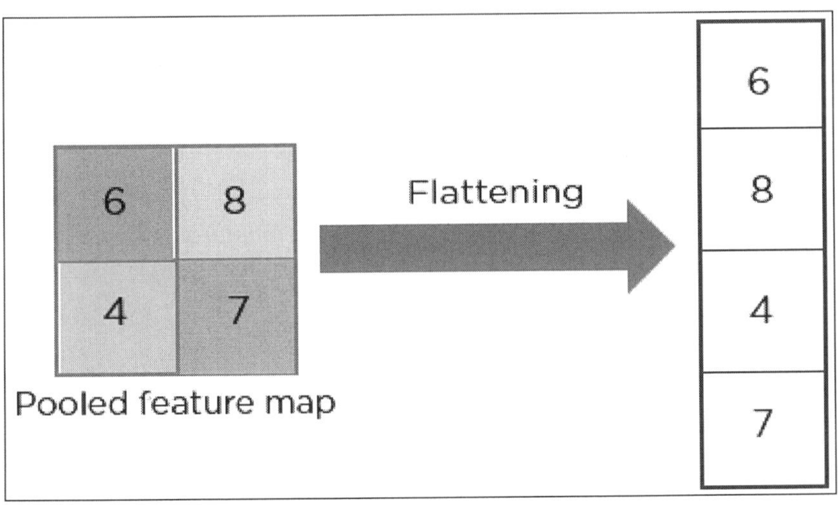

The flattened matrix is fed as input to the fully connected layer to classify the image.

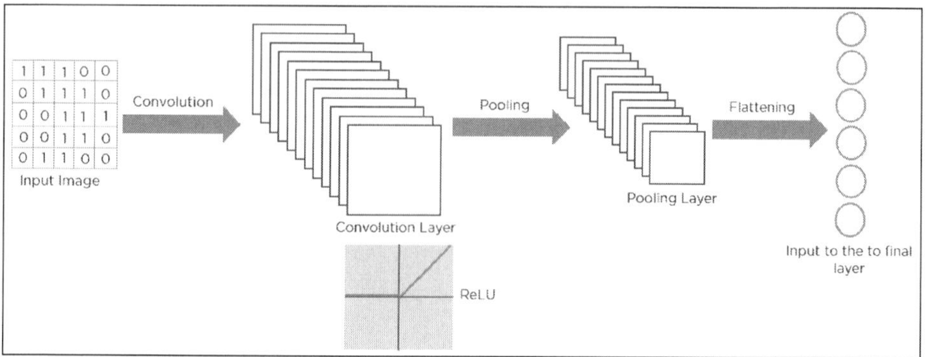

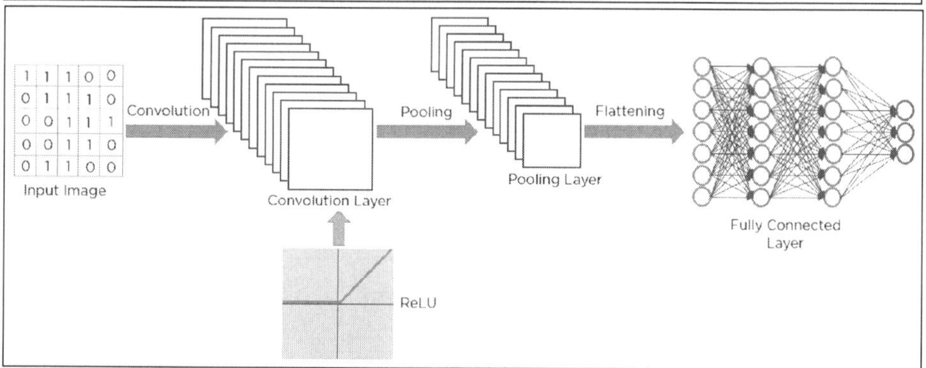

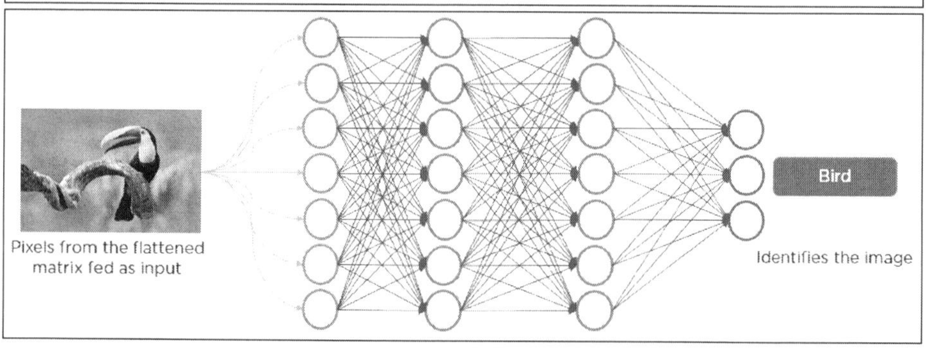

Here's how exactly CNN recognizes a bird:

- The pixels from the image are fed to the convolutional layer that performs the convolution operation
- It results in a convolved map
- The convolved map is applied to a ReLU function to generate a rectified feature map
- The image is processed with multiple convolutions and ReLU layers for locating the features

☆ Different pooling layers with various filters are used to identify specific parts of the image

☆ The pooled feature map is flattened and fed to a fully connected layer to get the final output

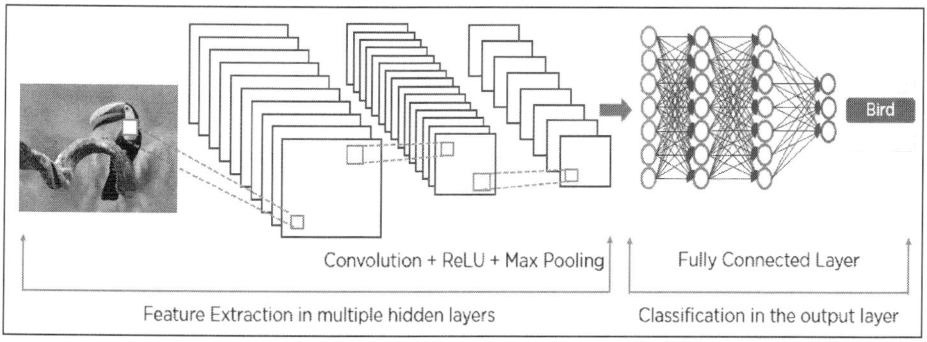

References

Alzubaidi, L., Zhang, J., Humaidi, A.J. *et al.*, Review of deep learning: concepts, CNN architectures, challenges, applications, future directions. *J Big Data* **8**, 53 (2021). https: //doi.org/10.1186/s40537-021-00444-8

https: //www.geeksforgeeks.org/introduction-convolution-neural-network/

Lakhani P, Sundaram B (2017) Deep learning at chest radiography: automated classification of pulmonary tuberculosis by using convolutional neural networks. Radiology 284: 574–582

S. Albawi, T. A. Mohammed and S. Al-Zawi, "Understanding of a convolutional neural network," 2017 International Conference on Engineering and Technology (ICET), Antalya, Turkey, 2017, pp. 1-6, doi: 10.1109/ICEngTechnol.2017.8308186. keywords: {Convolution;Neurons;Convolutional neural networks;Feature extraction;Image edge detection;machine learning;artificial neural networks;deep learning;convolutional neural networks;computer vision;Image recognition},

Yamashita, R., Nishio, M., Do, R.K.G. *et al.,* Convolutional neural networks: an overview and application in radiology. *Insights Imaging* **9**, 611–629 (2018). https: //doi.org/10.1007/s13244-018-0639-9

Yasaka K, Akai H, Abe O, Kiryu S (2018) Deep learning with convolutional neural network for differentiation of liver masses at dynamic contrast-enhanced CT: a preliminary study. Radiology 286: 887–896

Chapter 9

Feature Engineering for Machine Learning

Feature engineering is the pre-processing step of machine learning, which is used to transform raw data into features that can be used for creating a predictive model using Machine learning or statistical Modelling. Feature engineering in machine learning aims to improve the performance of models.

What is a Feature?

Generally, all machine learning algorithms take input data to generate the output. The input data remains in a tabular form consisting of rows (instances or observations) and columns (variable or attributes), and these attributes are often known as features. For example, an image is an instance in computer vision, but a line in the image could be the feature. Similarly, in NLP, a document can be an observation, and the word count could be the feature. So, we can say a feature is an attribute that impacts a problem or is useful for the problem.

What is Feature Engineering?

Feature engineering is the pre-processing step of machine learning, which extracts features from raw data. It helps to represent an underlying problem to predictive models in a better way, which as a result, improve the accuracy of the model for unseen data. The predictive model contains predictor variables and an outcome variable, and while the feature engineering process selects the most useful predictor variables for the model.

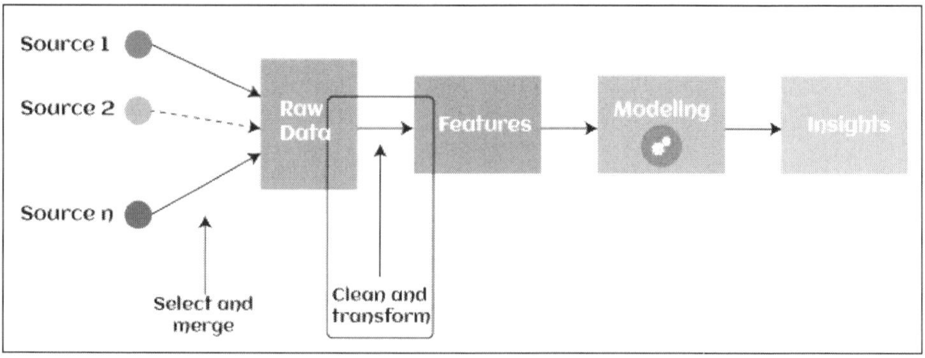

Since 2016, automated feature engineering is also used in different machine learning software that helps in automatically extracting features from raw data. Feature engineering in ML contains mainly four processes: Feature Creation, Transformations, Feature Extraction, and Feature Selection.

These processes are described as below:

Feature Creation: Feature creation is finding the most useful variables to be used in a predictive model. The process is subjective, and it requires human creativity and intervention. The new features are created by mixing existing features using addition, subtraction, and ration, and these new features have great flexibility.

Transformations: The transformation step of feature engineering involves adjusting the predictor variable to improve the accuracy and performance of the model. For example, it ensures that the model is flexible to take input of the variety of data; it ensures that all the variables are on the same scale, making the model easier to understand. It improves the model's accuracy and ensures that all the features are within the acceptable range to avoid any computational error.

Feature Extraction: Feature extraction is an automated feature engineering process that generates new variables by extracting them from the raw data. The main aim of this step is to reduce the volume of data so that it can be easily used and managed for data modelling. Feature extraction methods include cluster analysis, text analytics, edge detection algorithms, and principal components analysis (PCA).

Feature Selection: While developing the machine learning model, only a few variables in the dataset are useful for building the model, and the rest features are either redundant or irrelevant. If we input the dataset with all these redundant and irrelevant features, it may negatively impact and reduce the overall performance and accuracy of the model. Hence it is very important to identify and select the most appropriate features from the data and remove the irrelevant or less important features, which is done with the help of feature selection in machine learning. "Feature selection is a way of selecting the subset of the most relevant features from the original features set by removing the redundant, irrelevant, or noisy features."

Below are some benefits of using feature selection in machine learning:

- ☆ It helps in avoiding the curse of dimensionality.
- ☆ It helps in the simplification of the model so that the researchers can easily interpret it.
- ☆ It reduces the training time.
- ☆ It reduces overfitting hence enhancing the generalization.
- ☆ Need for Feature Engineering in Machine Learning

In machine learning, the performance of the model depends on data pre-processing and data handling. But if we create a model without pre-processing or data handling, then it may not give good accuracy. Whereas, if we apply feature engineering on the same model, then the accuracy of the model is enhanced. Hence, feature engineering in machine learning improves the model's performance. Below are some points that explain the need for feature engineering:

Better features mean flexibility.

In machine learning, we always try to choose the optimal model to get good results. However, sometimes after choosing the wrong model, still, we can get better predictions, and this is because of better features. The flexibility in features will enable you to select the less complex models. Because less complex models are faster to run, easier to understand and maintain, which is always desirable.

Better features mean simpler models.

If we input the well-engineered features to our model, then even after selecting the wrong parameters (Not much optimal), we can have good outcomes. After feature engineering, it is not necessary to do hard for picking the right model with the most optimized parameters. If we have good features, we can better represent the complete data and use it to best characterize the given problem.

Better features mean better results.

As already discussed, in machine learning, as data we will provide will get the same output. So, to obtain better results, we must need to use better features.

Steps in Feature Engineering

The steps of feature engineering may vary as per different data scientists and ML engineers. However, there are some common steps that are involved in most machine learning algorithms, and these steps are as follows:

Data Preparation: The first step is data preparation. In this step, raw data acquired from different resources are prepared to make it in a suitable format so that it can be used in the ML model. The data preparation may contain cleaning of data, delivery, data augmentation, fusion, ingestion, or loading.

Exploratory Analysis: Exploratory analysis or Exploratory data analysis (EDA) is an important step of features engineering, which is mainly used by data scientists. This step involves analysis, investing data set, and summarization of the main characteristics of data. Different data visualization techniques are used to better understand the manipulation of data sources, to find the most appropriate statistical technique for data analysis, and to select the best features for the data.

Benchmark: Benchmarking is a process of setting a standard baseline for accuracy to compare all the variables from this baseline. The benchmarking process is used to improve the predictability of the model and reduce the error rate.

Feature Engineering Techniques

Some of the popular feature engineering techniques include:

1. Imputation

Feature engineering deals with inappropriate data, missing values, human interruption, general errors, insufficient data sources, *etc.* Missing values within the dataset highly affect the performance of the algorithm, and to deal with them "Imputation" technique is used. Imputation is responsible for handling irregularities within the dataset.

For example, removing the missing values from the complete row or complete column by a huge percentage of missing values. But at the same time, to maintain the data size, it is required to impute the missing data, which can be done as:

For numerical data imputation, a default value can be imputed in a column, and missing values can be filled with means or medians of the columns.

For categorical data imputation, missing values can be interchanged with the maximum occurred value in a column.

2. Handling Outliers

Outliers are the deviated values or data points that are observed too away from other data points in such a way that they badly affect the performance of the model. Outliers can be handled with this feature engineering technique. This technique first identifies the outliers and then remove them out.

Standard deviation can be used to identify the outliers. For example, each value within a space has a definite to an average distance, but if a value is greater distant than a certain value, it can be considered as an outlier. Z-score can also be used to detect outliers.

3. Log Transform

Logarithm transformation or log transform is one of the commonly used mathematical techniques in machine learning. Log transform helps in handling the skewed data, and it makes the distribution more approximate to normal after transformation. It also reduces the effects of outliers on the data, as because of the normalization of magnitude differences, a model becomes much robust.

4. Binning

In machine learning, overfitting is one of the main issues that degrade the performance of the model and which occurs due to a greater number of parameters and noisy data. However, one of the popular techniques of feature engineering, "binning", can be used to normalize the noisy data. This process involves segmenting different features into bins.

5. Feature Split

As the name suggests, feature split is the process of splitting features intimately into two or more parts and performing to make new features. This technique helps the algorithms to better understand and learn the patterns in the dataset.

The feature splitting process enables the new features to be clustered and binned, which results in extracting useful information and improving the performance of the data models.

6. One Hot Encoding

One hot encoding is the popular encoding technique in machine learning. It is a technique that converts the categorical data in a form so that they can be easily understood by machine learning algorithms and hence can make a good prediction. It enables group the of categorical data without losing any information.

References

Akaike, Hirotugu. 1974. "A New Look at the Statistical Model Identification." *IEEE Transactions on Automatic Control* 19 (6). Ieee: 716–23.

Galli, S. 2020. Python Feature Engineering Cookbook: Over 70 Recipes for Creating, Engineering, and Transforming Features to Build Machine Learning Models. Packt Publishing. https: //books.google.com/books?id=2c_LDwAAQBAJ.

https: //www.geeksforgeeks.org/what-is-feature-engineering/

https: //www.ibm.com/topics/feature-engineering

https: //www.javatpoint.com/feature-engineering-for-machine-learning

J. Heaton, "An empirical analysis of feature engineering for predictive modeling," SoutheastCon 2016, Norfolk, VA, USA, 2016, pp. 1-6, doi: 10.1109/ SECON.2016.7506650. keywords: {Mathematical model;Biological neural networks;Neurons;Linear regression;Support vector machines;Analytical models},

Rawat, Tara and Khemchandani, Vineeta. (2019). Feature Engineering (FE) Tools and Techniqucs for Better Classification Performance. 10.21172/ijiet.82.024.

Chapter 10

Demonstration Using MNIST Dataset

What is the MNIST Dataset?

https: //www.digitalocean.com/community/tutorials/mnist-dataset-in-python - what-is-the-mnist-dataset

MNIST set is a large collection of handwritten digits. It is a very popular dataset in the field of image processing. It is often used for benchmarking machine learning algorithms.

MNIST is short for **Modified National Institute of Standards and Technology database.**

MNIST contains a collection of **70,000, 28 x 28** images of handwritten digits from **0 to 9.**

The dataset is already divided into training and testing sets. We will see this later in the tutorial.

For more information on MNIST, refer to its Wikipedia page. We are going to import the dataset from Keras.

Let's start with loading the dataset into our python notebook.

Loading MNIST from Keras

https: //www.digitalocean.com/community/tutorials/mnist-dataset-in-python - loading-mnist-from-keras

We will first have to import the MNIST dataset from the **Keras module**.

We can do that using the following line of code:

```
from keras.datasets import mnist
```

Now we will load the training and testing sets into separate variables.

```
(train_X, train_y), (test_X, test_y) = mnist.load_data()
```

Let's find out how many images are there in the training and testing sets. In other words, let's try and find out the split ratio of the this dataset.

To learn more about split ratio, refer to this tutorial on how to slit data into training and testing sets.

To find the split ratio we are going to print the shapes of all the vectors.

```
print('X_train: ' + str(train_X.shape))print('Y_train: ' + str(train_y.shape))
print('X_test: ' + str(test_X.shape))print('Y_test: ' + str(test_y.shape))
```

Output:

X_train: (60000, 28, 28)

Y_train: (60000,)

X_test: (10000, 28, 28)

Y_test: (10000,)

We can see that there are **60k images in the training set** and **10k images in the testing set.**

The dimension of our training vector is **(60000, 28, 28)**, this is because there are **60,000 grayscale images** with the dimension **28X28.**

Complete Code for Loading the MNIST Dataset

https: //www.digitalocean.com/community/tutorials/mnist-dataset-in-python - complete-code-for-loading-the-mnist-dataset

Here's the complete code from this section :

```
from keras.datasets import mnist
#loading the dataset(train_X, train_y), (test_X, test_y) = mnist.load_data()
```

#printing the shapes of the vectors print('X_train: ' + str(train_X.shape)) print('Y_train: ' + str(train_y.shape))print('X_test: ' + str(test_X.shape))print('Y_test: ' + str(test_y.shape))

Let's learn how to plot this dataset.

Plotting the MNIST Dataset using Matplotlib

https: //www.digitalocean.com/community/tutorials/mnist-dataset-in-python - plotting-the-mnist-dataset-using-matplotlib

It is always a good idea to plot the dataset you are working on. It will give you a good idea about the kind of data you are dealing with.

As a responsible data scientist, it should be your duty to always plot the dataset as step zero.

To plot the dataset, use the following piece of code :

from matplotlib import pyplotfor i in range(9):

pyplot.subplot(330 + 1 + i)

pyplot.imshow(train_X[i], cmap=pyplot.get_cmap('gray'))

pyplot.show()

Output :

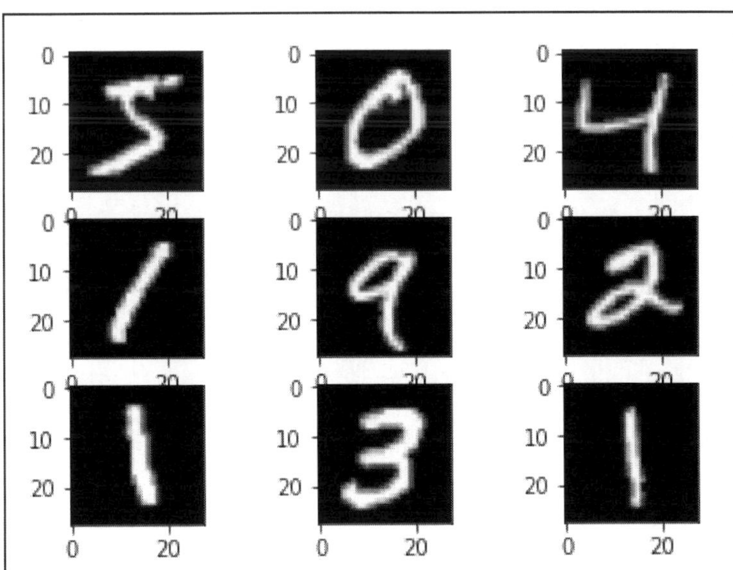

MNIST Data Plotted

References

Cheng-Lin Liu, K. Nakashima, H. Sako and H. Fujisawa, "Handwritten digit recognition using state-of-the-art techniques," Proceedings Eighth International Workshop on Frontiers in Handwriting Recognition, Niagra-on-the-Lake, ON, Canada, 2002, pp. 320-325, doi: 10.1109/IWFHR.2002.1030930. keywords: {Handwriting recognition;Computed tomography;Chromium;Conferences}

Gope, Birjit and Pande, Sagar and Karale, Nikhil and Dharmale, Shivani and Umekar, Pooja. (2021). Handwritten Digits Identification Using Mnist Database Via Machine Learning Models Handwritten Digits Identification Using Mnist Database Via Machine Learning Models. IOP Conference Series: Materials Science and Engineering. 1022. 012108. 10.1088/1757-899X/1022/1/012108.

https: //towardsdatascience.com/going-beyond-99-mnist-handwritten-digits-recognition-cfff96337392

L. Deng, "The MNIST Database of Handwritten Digit Images for Machine Learning Research [Best of the Web]," in IEEE Signal Processing Magazine, vol. 29, no. 6, pp. 141-142, Nov. 2012, doi: 10.1109/MSP.2012.2211477.

Chapter 11

EMD Based Hybrid Models for Time Series Forecasting

Introduction to Time Series Forecasting

A **time series** is a sequence of data points collected at successive equally spaced intervals of time (hourly, daily, weekly, monthly, quarterly, *etc.*). Each data point typically represents a measurement or observation taken at a specific instant in time. Time series data is characterized by its temporal ordering, where the order of observations is critical due to dependencies over time. Key characteristics of time series data include:

- ☆ **Temporal Dependence**: The value of a data point depends on previous values.
- ☆ **Seasonality**: Patterns that repeat at predictable intervals.
- ☆ **Trend**: Long-term increase or decrease in data over time.
- ☆ **Noise**: Random fluctuations that make prediction challenging.

Time series forecasting is crucial in various fields for predicting future values based on past observations. In economics and finance, it is used to forecast stock prices, commodity prices, and economic indicators, as well as to predict the demand for products and services. In business, it aids in sales forecasting to optimize inventory management and production schedules and in predicting customer behaviour for marketing strategies. In weather and climate studies, it helps forecast temperature, precipitation, and other meteorological variables, and predict climate patterns for agriculture and disaster management. In healthcare, it is valuable for predicting patient admissions, disease outbreaks, and medical resource demand. In engineering, it is used to predict equipment failure and maintenance schedules and to forecast energy consumption and resource usage. In the social sciences, it helps in predicting population growth, crime rates, and census data.

Traditional Techniques for Time Series Forecasting

The various time series approaches developed for forecasting can be broadly divided into two categories, *i.e.*, statistical models and artificial intelligence (AI) models. Some statistical models are briefly discussed below:

A. Moving Average Methods

Moving Average (MA) methods are simple yet effective for smoothing time series data and making short-term predictions. It computes the average of a subset of recent data points to forecast the next point. It is of broadly three types:

- **Simple Moving Average (SMA)**: Equal weights to all data points in the window.
- **Weighted Moving Average (WMA)**: Assigns different weights to different data points.
- **Exponential Moving Average (EMA)**: Gives more weight to recent observations using an exponentially decreasing weight.

B. Exponential Smoothing

Exponential Smoothing methods are used for time series forecasting by assigning exponentially decreasing weights over time. It is useful for capturing trends and seasonality in data:

- **Single Exponential Smoothing**: Smoothes out random fluctuations in data.
- **Double Exponential Smoothing (Holt's method)**: Accounts for trends in addition to smoothing.
- **Triple Exponential Smoothing (Holt-Winters method)**: Incorporates seasonality along with trend and level components.

C. Autoregressive Integrated Moving Average (ARIMA)

A univariate time series can be modelled by expressing it as a function of its own lagged or past values and some random disturbances occurring in it. The ARIMA (*p*, *d*, *q*) model with *p*, *d*, and *q* as orders of autoregression, differencing (integration), and moving average, respectively can be expressed as:

$$\varphi(B)\,\Delta^{d}y_t = \theta(B)u_t$$

where y_t is the value of time series at time t, u_t is the disturbance term at time t which is assumed to be random and identically distributed with mean zero and constant variance σ_2, the backshift operator B is defined by $By_t = y_t{-}1$, $\Delta = (1 - B)$

is the differencing operator, $\varphi(B)$ and $\theta(B)$ are the polynomials of degree p and q in B respectively. Box-Jenkins methodology is a systematic approach to identify, fit, and forecast ARIMA models.

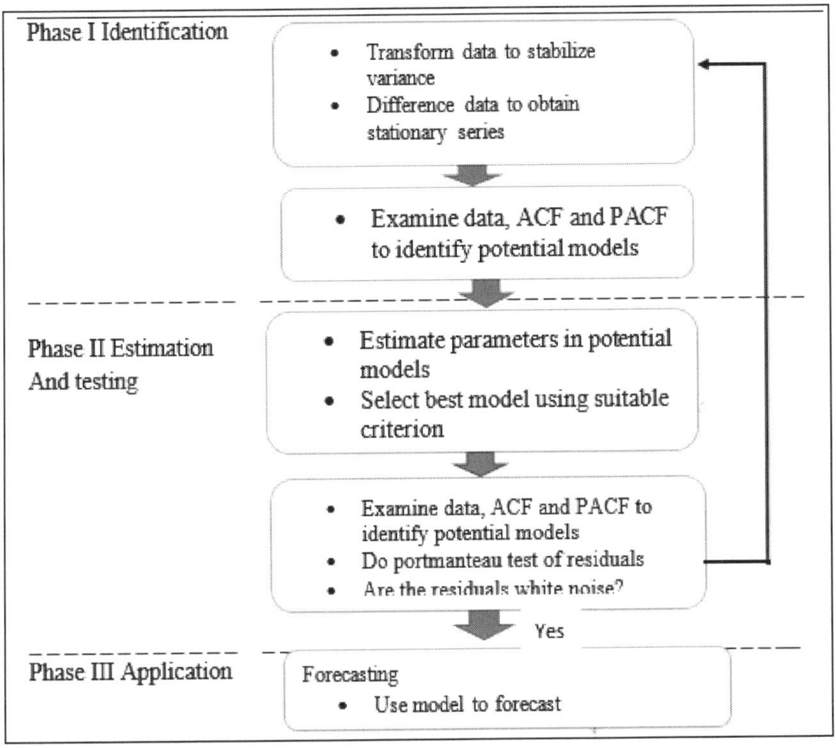

Summary of Box-Jenkins Methodology for ARIMA Modelling.

Challenges with Traditional Techniques

A. Handling Non-linearity and Complex Patterns

Traditional time series forecasting techniques often struggle with non-linear relationships and complex patterns within data:

- ✰ **Non-linear Relationships**: Many real-world phenomena exhibit non-linear behaviour, which classical methods like ARIMA may not adequately capture. Thus, several suitable nonlinear time series models such as the regime-switching models, like selfexciting threshold autoregressive (SETAR), smooth transition autoregressive (STAR), *etc.* were suggested for forecasting, but these models too, require the specific nonlinear relations among data and lack generalization ability.
- ✰ **Complex Patterns**: Time series data can contain intricate patterns such as abrupt changes, seasonality with irregular frequencies, and interactions between multiple variables.

B. Long-term Dependencies

Traditional methods like ARIMA and exponential smoothing cannot effectively capture long term dependencies in time series data:

- **Short Memory**: These models typically rely on recent observations and have limited memory of historical data.
- **Inability to Learn Long-term Patterns**: Time series with long-term trends or seasonal patterns over extended periods may require more sophisticated approaches to model effectively.

C. Limited Ability to Incorporate External Factors

Classical techniques often struggle to integrate external factors or predictors into the forecasting process:

- **Exogenous Variables**: Factors external to the time series data, such as economic indicators or weather conditions, can significantly influence future trends.
- **Static Relationships**: Traditional methods assume stationary relationships and may not adapt well to changes in external factors over time.

Artificial Intelligence (AI) Models

The term "artificial intelligence" dates back to 1955 when John McCarthy coined it and defined it as "the science and engineering of making intelligent machines". In general, AI encompasses techniques that mimic the human brain. The term "machine learning" is a subfield of AI, where a machine learns and improves from experience without any explicit programming. Machine learning (ML) does this through various algorithms like artificial neural networks, support vector machines, decision trees, and genetic algorithms. ML models, with their flexible operational designs and potent self-learning abilities, have recently turned out to be a promising alternative for time series forecasting. Artificial neural networks (ANNs), one of the important areas of ML, are data-driven and follow a self-adaptive approach, unlike conventional modelbased methods. They are powerful tools for modelling especially when the underlying datagenerating process is not known. They imitate the learning process of the biological neural network (brain) through some simple computational units that are highly interconnected called nodes or neurons. The most important part of these neurons is activation function. An activation function is a mathematical function that converts the input to output and adds the magic of neural network processing. Activation functions give the nonlinearity property to neural networks and make them true universal function approximators. Without activation functions, a neural network would essentially act as a linear regression model, regardless of its depth. There are several commonly used activation functions, each with its unique properties and applications. The sigmoid function maps input values to an output range between 0 and 1, making it useful for models where the output needs to be interpreted as probabilities. The hyperbolic tangent (tanh) function, which outputs values between -1 and 1, is often preferred over the sigmoid function in hidden layers as it generally performs better in practice. The Rectified Linear Unit (ReLU) function is widely used due to its simplicity and effectiveness in mitigating the vanishing gradient problem by outputting zero for negative inputs and a linear relationship for positive inputs. The choice of activation function can significantly impact the training dynamics and performance of a neural network, making it a critical decision in the design of ANN architectures. ANNs are being applied in almost every discipline like genetics, statistics, biotechnology, finance, industry, *etc.* for solving problems like classification, regression, prediction, *etc.* Typically, a neural network can be used effectively for pattern recognition, but mainly for unstructured static data that is not time-constrained. However, it is interesting to see how a neural network can be employed for temporal pattern recognition. The difference between the two types is that in the second case, the patterns evolve over time, which makes recognition harder. In consequence, the neural network has to adapt to these changes or at least consider them in the learning process.

Time series prediction is a perfect example of a temporal changing pattern. For temporal processing, we build time into the operation of a neural network mainly in two ways. One way is to convert a "static" neural network into a dynamic one using a memory structure. Shortterm and long-term memory are two types of memory. Long-term memory is built into a neural network through supervised learning, whereby the information content of the training data set is stored in the synaptic weights of the network. However, if the task at hand has a temporal dimension, short-term memory is built through the use of time delays at the synaptic level inside the network or the input layer of the network (Jha and Sinha, 2014). Another important way for temporal processing to occur is through the use of feedback. Neural networks with one or more feedback loops are referred to as "recurrent neural networks (RNNs)". Thus, we have two candidates for temporal processing: 1) time-lagged feedforward neural networks, and 2) recurrent neural networks.

A. Time Delay Neural Network (TDNN) Model

TDNN is a type of feed-forward neural network model that is being used for time series forecasting successfully (Jha and Sinha, 2014). This neural network model builds a short-term memory in its network through the use of time delays of a univariate time series to capture the temporal dimension of the series (Haykin, 2010).

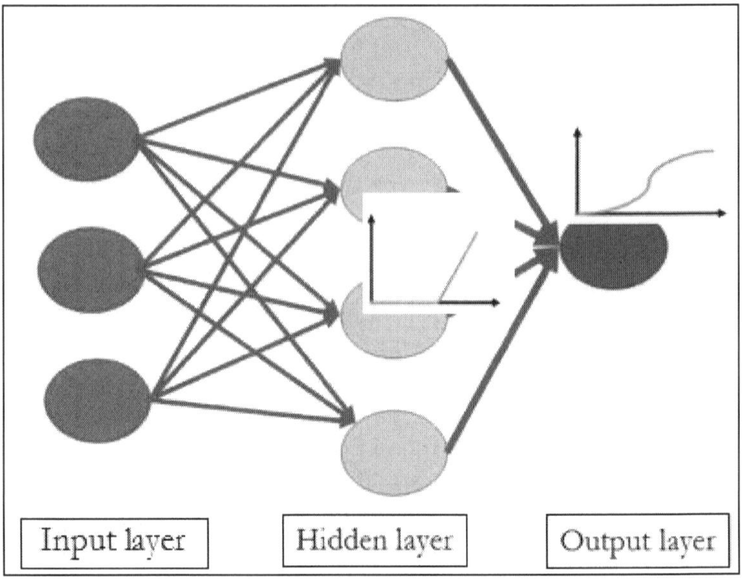

Feed-Forward Neural Network: Information Propagates in the Forward Direction *i.e.* from Input to Output.

To achieve this, a time series is first converted into a supervised learning format as a collection of samples. Each sample constitutes an input component (***X***) and an output component (***Y***). For example, if a time series contain N observations $y_1, y_2, y_3, ..., y_N$ and the model is to be made using p-lagged values as input nodes, then to get one step ahead prediction the first sample will contain $y_1, y_2, y_3, ..., y_p$ as input components and y_{p+1} will be the output component. In this way, a set of N-p samples are generated each consisting input vector (***X***) and an output variable (***Y***). These samples are fed into an algorithm to learn the mapping function from the input to the output *i.e.* $Y = f(\boldsymbol{X})$.

The task of the algorithm is to approximate or learn the real underlying mapping function so well that it can predict the output variables with maximum possible accuracy whenever new input data is given to it. The whole process of learning is called the training of the neural network. The training consists of several iterations of the propagation of signals in both forward and backward directions. In forward propagation, the example sets having input and output values obtained using the supervised learning format are fed into the networks. The number of input values of the samples determines the number of input nodes in the network. The input values are then passed to the nodes of the hidden layer as their weighted sum where some affine transformations are done by nonlinear activation functions of the hidden nodes. Mathematically, the forward propagation can be expressed as

$$y_{t+1} = \gamma\left(\sum_{m=1}^{q} \partial_m \Psi\left(\sum_{j=0}^{p} w_{mj} y_{t-j} + b_m\right)\right)$$

nodes, respectively, and is the output of the neuron at *the* time step. In backward propagation, a loss function or cost function is computed by comparing the model output and desired output. This loss function is traversed backward in the network computing its gradient or partial derivative with respect to all the weights in a particular order as a chain rule. The partial derivative with negative direction accounts for gradient descent which updates the synaptic weights of the network intending to minimize the loss function as small as possible. These several iterations of forward and backward propagation of information of all the examples or samples are called an epoch of the training. Thus, the training of the model requires the optimization of the number of epochs along with several other hyperparameters like the number of input nodes, the number of hidden nodes, batch size, *etc.* Tuning of these hyperparameters is problem dependent and determined through experimentation and grid search methods on the given data.

Empirical Mode Decomposition (EMD)

A review of literature suggest that neural network models are not able to model a nonstationary time series data (Nelson *et al.*, 1999). Huang *et al.* (1998) proposed the empirical mode decomposition (EMD) method for extracting characteristic information from non-linear and non-stationary time series through a divide and conquer concept. It is an adaptive decomposition with which any complicated series can be decomposed into its intrinsic mode functions (IMFs).

Method

- **Identification of Local Extrema**: The algorithm identifies all the local maxima and minima in the time series.
- **Envelope Creation**: Two envelopes are created—one connecting all the local maxima and the other connecting all the local minima using cubic spline interpolation.
- **Mean Calculation**: The mean of these two envelopes is computed.
- **Extraction of IMF**: The mean is subtracted from the original time series to obtain the first IMF. This IMF should have zero mean and exhibit symmetric upper and lower envelopes.
- **Iteration**: The residual obtained by subtracting the IMF from the original series is treated as the new series, and the process is repeated to extract subsequent IMFs until the residual becomes a monotonic function or a function with only one extremum.

After full decomposition we can get original series in this form

$$
\begin{aligned}
y_t &= c_t(1) + r_t(1) \\
&= c_t(1) + c_t(2) + r_t(2) \\
&= c_t(1) + c_t(2) + c_t(3) + r_t(3) \\
&\vdots \\
&= \textstyle\sum_{j=1}^{N} c_t(j) + r_t(n)
\end{aligned}
$$

i.e. several IMFs (each representing different frequency components of the original time series) and one residual component are obtained. Here, total no. of IMF = $log2T$, where T is length of data series.

$$\hat{y}(t) = \sum_{j=1}^{n} \hat{c}_j(t) + \hat{r}(t)$$

B. EMD-ARIMA Model

On the similar line to EMD-TDNN, EMD-ARIMA is a hybrid approach that leverages the strengths of Empirical Mode Decomposition (EMD) and Auto-Regressive Integrated Moving Average (ARIMA) to effectively forecast complex time series data. Choosing between EMDARIMA and EMD-TDNN for time series forecasting depends on several key factors, including the nature of the data, complexity of the time series, and specific forecasting requirements. EMD-ARIMA is advantageous for modelling linear relationships within the decomposed Intrinsic Mode Functions (IMFs). It is particularly effective when the IMFs exhibit linear characteristics, as ARIMA can efficiently capture these linear patterns. Additionally, ARIMA offers interpretability, allowing for clear insights into the autoregressive and moving average components of the data, which can be beneficial for understanding underlying trends. This model is also well-documented and straightforward to implement, making it suitable for scenarios where a simpler, more interpretable model is preferred, and computational resources are limited. EMD-ARIMA is ideal when the time series data is mostly linear postdecomposition, shows clear seasonal or cyclic patterns, and when the IMFs are stationary or can be easily transformed to stationary series.

On the other hand, EMD-TDNN excels in capturing complex non-linear relationships within the IMFs. Time Delay Neural Networks (TDNN) can learn intricate patterns through their hidden layers, making them suitable for modelling subtle and complex dependencies in the data. This flexibility allows TDNN to adapt to a wide range of data characteristics and complexities, leveraging parallel processing capabilities for handling large datasets and realtime applications. EMD-TDNN is preferable when the IMFs exhibit non-linear patterns and dynamic changes, requiring a more complex and flexible model. This approach is especially useful when sufficient computational resources are available to train and deploy neural network models.

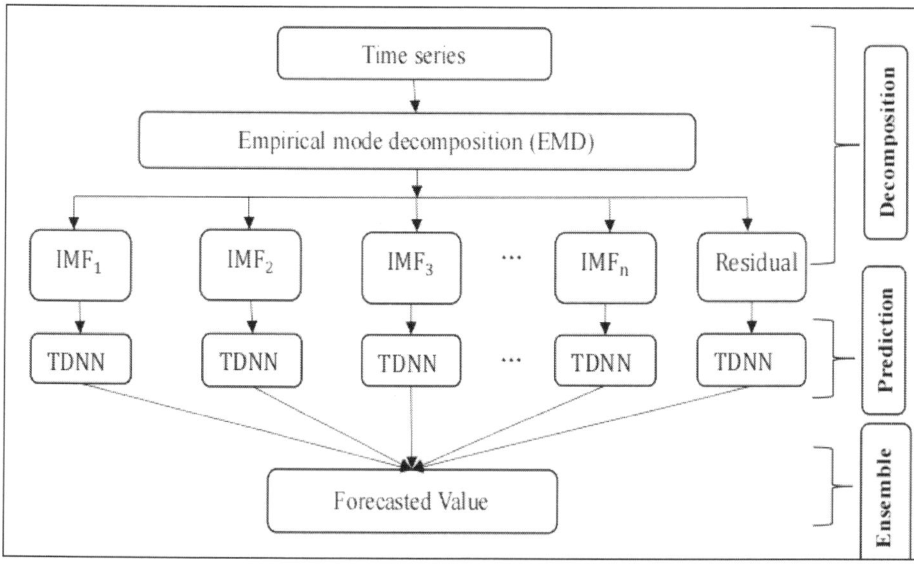

Flowchart of EMD-TDNN Model for Time Series Forecasting.

Conclusion

In this chapter, we have explored the aspects of time series forecasting using advanced hybrid models, specifically EMD-ARIMA and EMD-TDNN. By understanding the decomposition of complex non-linear and non-stationary time series into simpler Intrinsic Mode Functions (IMFs) using Empirical Mode Decomposition (EMD), we can leverage the strengths of both ARIMA and neural network models to achieve more accurate and reliable forecasts. The choice between EMD-ARIMA and EMD-TDNN hinges on the nature of the data, with EMD-ARIMA being more suitable for linear and stationary IMFs, and EMD-TDNN excelling in capturing non-linear and dynamic patterns. This comprehensive understanding and application of hybrid time series models equip us with the capability to make informed decisions based on robust and precise predictions.

References

Choudhary, K., Jha, G. K., Jaiswal, R., Mishra, D. C., and Chaturvedi, K. K. (2022). Decomposition-based auto-regressive model for agricultural price prediction. *Indian Journal of Extension Education*, 58(2), 83-87.

Haykin, S. (2010). *Neural Networks and Learning Machines*. PHI Learning.

Huang, N. E., Shen, Z., Long, S. R., Wu, M. C., Shih, H. H., Zheng, Q., and Liu, H. H. (1998). The empirical mode decomposition and the Hilbert spectrum for nonlinear and non stationary time series analysis. In *Proceedings of the Royal Society of London A: mathematical, physical and engineering sciences.* 454, 903-995.

Jaiswal, R., Jha, G. K., Kumar, R. R., and Choudhary, K. (2021). Deep long short-term memory based model for agricultural price forecasting. *Neural Computing and Applications.* https: //doi.org/10.1007/s00521-021-06621-3

Jha, G. K., and Sinha, K. (2014). Time-delay neural networks for time series prediction: An application to the monthly wholesale price of oilseeds in India. *Neural Computing and Applications, 24*(3–4), 563–571. https: //doi.org/10.1007/s00521-012-1264-z

LeCun, Y., Bengio, Y., and Hinton, G. (2015). Deep learning. *Nature, 521*(7553), 436–444. https: //doi.org/10.1038/nature14539

Chapter 12

Installation of R and R-Studio

To get and install R and RStudio on your computer, you can follow these general steps:

1. Visit the official R Project website: https: //www.r-project.org/

2. Click on the "CRAN" link on the left sidebar. CRAN stands for Comprehensive R Archive Network, which is the network of servers around the world that store R packages and the R software itself.

3. Choose a CRAN mirror near your location. This will redirect you to a mirror site where you can download R. (INDIA)

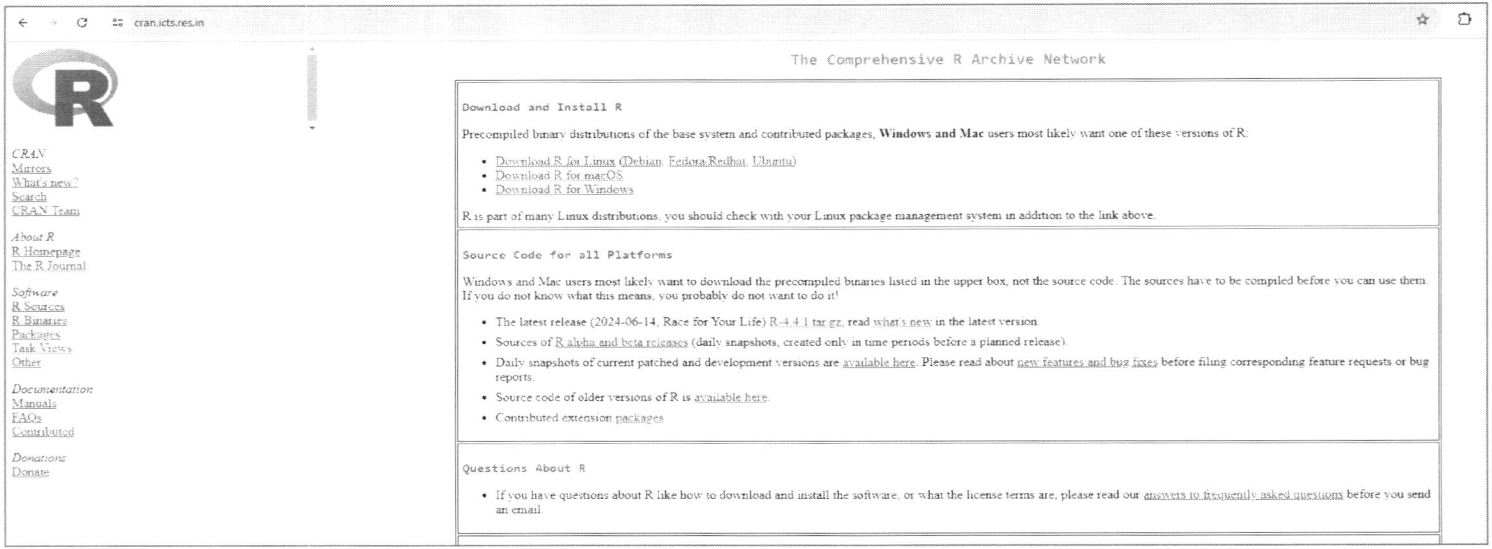

4. On the mirror site, select your operating system (*e.g.*, Windows, macOS, Linux) and click on the corresponding link to download the R installer.

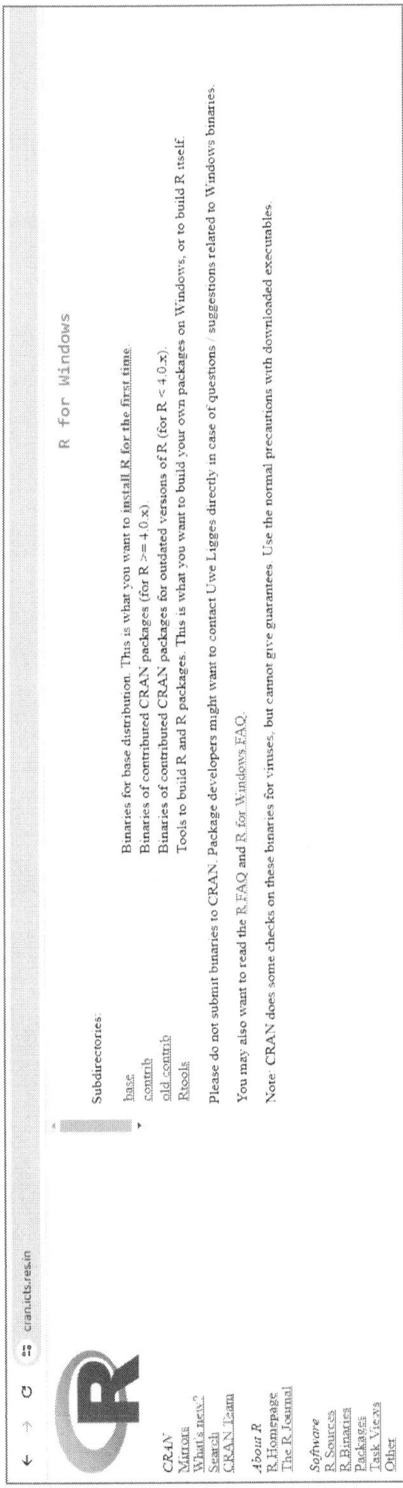

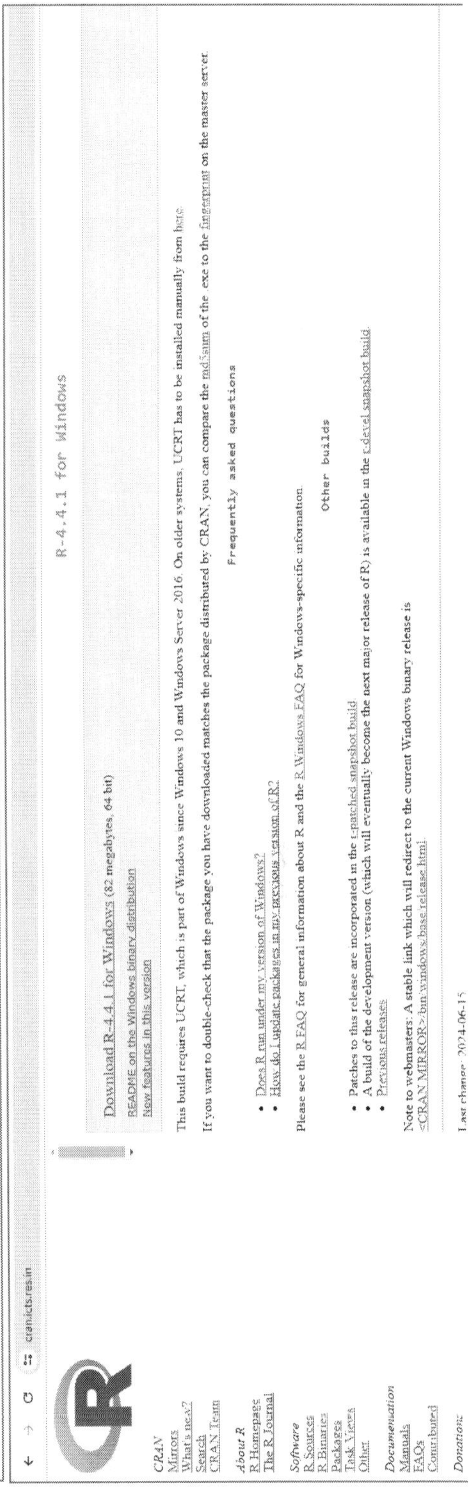

5. Once the download is complete, run the installer and follow the installation instructions. The default settings should work for most users, but you can customize the installation if desired.

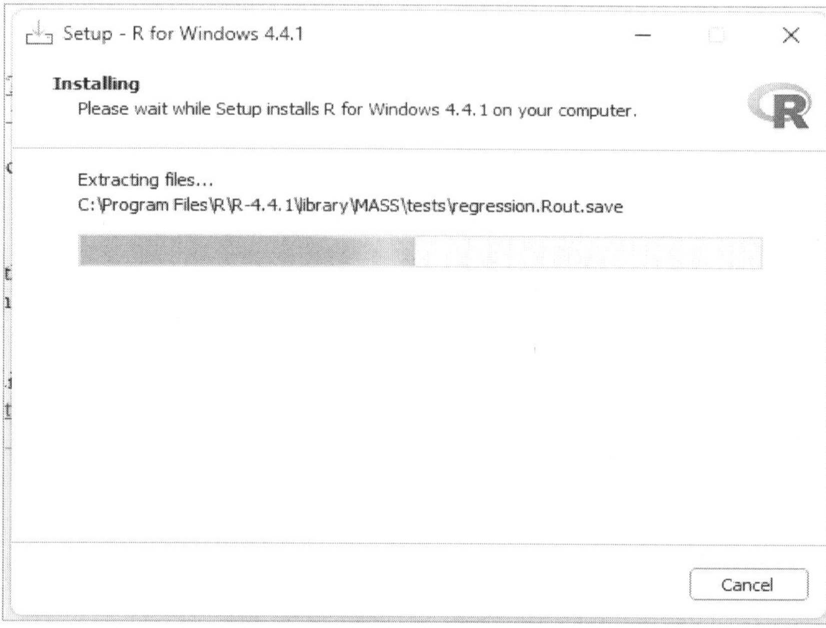

6. After the installation is complete, you should be able to launch R from your computer's applications or programs menu.

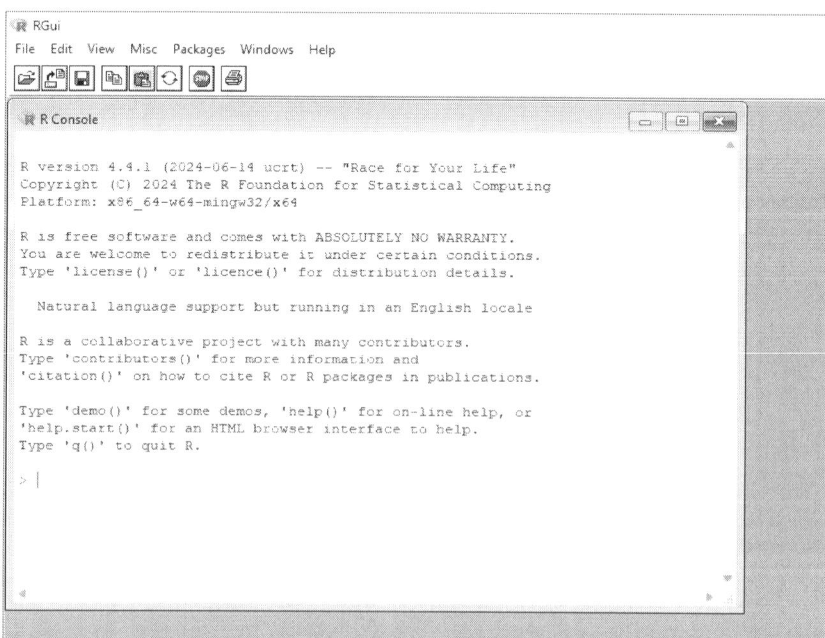

7. After R. you have to install an integrated development environment (IDE) for R, *i.e.* RStudio. RStudio provides a more user-friendly interface and additional features for working with R. You can download RStudio from the following website: https: //www.rstudio.com/products/rstudio/download/

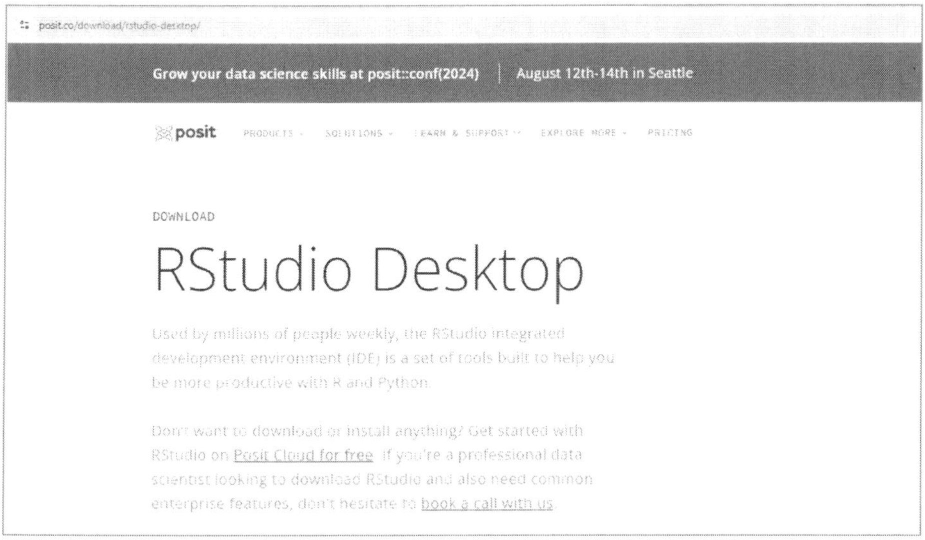

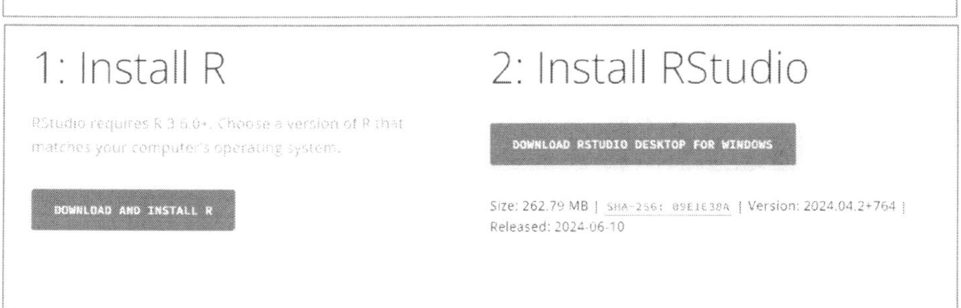

8. Follow the instructions to install RStudio on your computer.

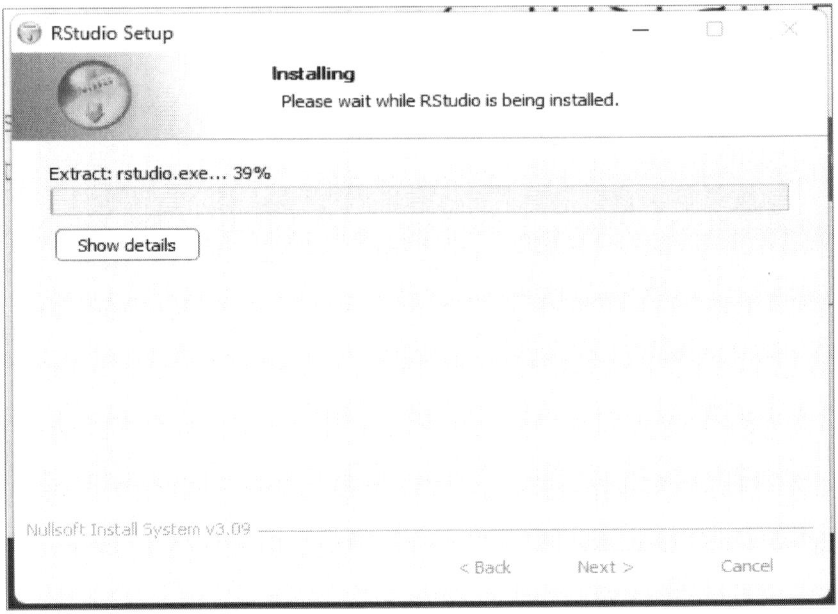

Once you have R and RStudio installed, you'll have the necessary tools to start working with the R programming language. You can launch RStudio and begin writing R code or analyzing data using R packages.

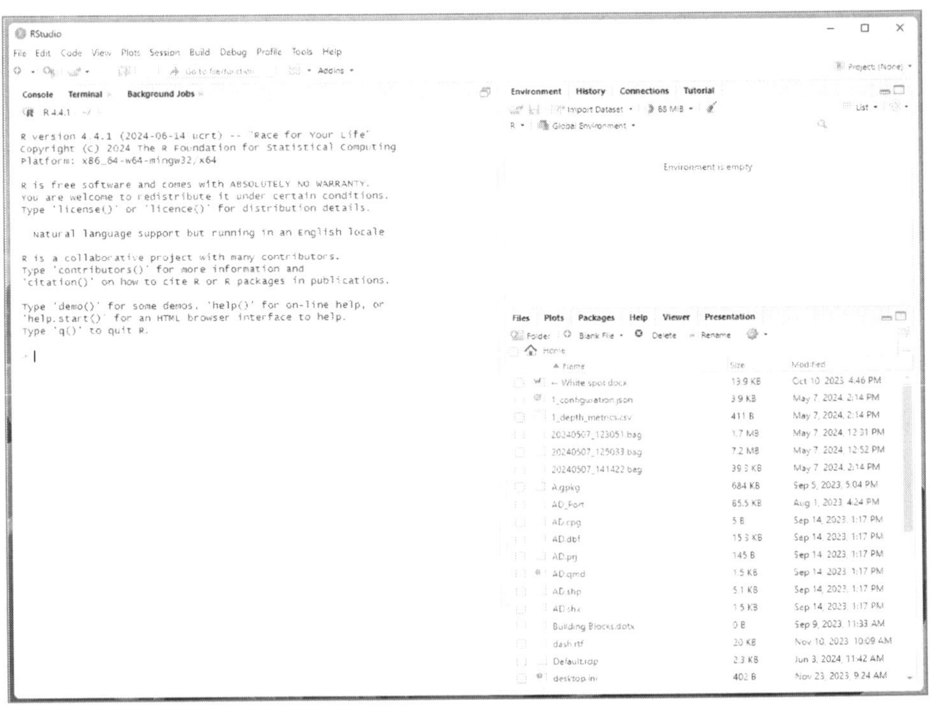

References

Hair, J.F., Hult, G.T.M., Ringle, C.M., Sarstedt, M., Danks, N.P., Ray, S. (2021). Overview of R and RStudio. In: Partial Least Squares Structural Equation Modeling (PLS-SEM) Using R. Classroom Companion: Business. Springer, Cham. https: //doi.org/10.1007/978-3-030-80519-7_2

Hair, Joseph and Hult, G. Tomas M. and Ringle, Christian and Sarstedt, Marko and Danks, Nicholas and Ray, Soumya. (2021). Overview of R and RStudio. 10.1007/978-3-030-80519-7_2.

https: //cran.r-project.org/bin/windows/base/

https: //posit.co/download/rstudio-desktop/

https: //rstudio-education.github.io/hopr/starting.html

https: //www.r-project.org/

Chapter 13

Practical on Classification and Prediction using Machine Learning Techniques

Mobile Price Range Classification and Prediction using Machine Learning Techniques

Mobile price classification is a common task in the field of machine learning where the goal is to build a predictive model that can classify mobile phones into different price ranges based on their features and specifications. The classification model aims to automate the process of determining the price category of a mobile phone, which can be useful for both consumers and sellers in making informed decisions. It provides valuable insights for consumers, helping them compare and make decisions based on estimated price ranges. For sellers, these models assist in determining competitive pricing strategies and understanding market dynamics.

Dataset

The dataset was initially retrieved from Kaggle under the title "Mobile Price Classification". A dataset is provided, and it is also provided into the repository. The dataset consists of a nearly 2000 data entries of mobile phone features (RAM, size, *etc.*), specifications *etc.* and their price range. The various features and information with each combination of those will result in different price range can be used to predict the price range of a mobile phone.

Methodology

This code demonstrates a complete machine learning pipeline:

- ✰ Loading and preparing data.
- ✰ Splitting data into training and testing sets.
- ✰ Initializing multiple classifiers.
- ✰ Training the classifiers.
- ✰ Evaluating the classifiers using accuracy.
- ✰ Taking user input to predict the mobile phone price range.
- ✰ Displaying the predictions and accuracies of the classifiers.

Step-by-Step Explanation of Program

1. Library Installation

You need to install the necessary libraries to run this code. The primary libraries required are pandas for data handling and scikit-learn for machine learning models and evaluation metrics.

To install these libraries, you can use the following commands:

```
pip install pandas scikit-learn
```

Ensure these libraries are installed in your Python environment before running the code.

2. Importing Libraries

```
import pandas as pd
from sklearn.model_selection import train_test_split
from sklearn.tree import DecisionTreeClassifier
from sklearn.svm import SVC
from sklearn.linear_model import LogisticRegression
from sklearn.ensemble import RandomForestClassifier
from sklearn.neighbors import KNeighborsClassifier
from sklearn.naive_bayes import GaussianNB
from sklearn.metrics import accuracy_score
```

- **pandas (pd)**: Used for data manipulation and reading the CSV file.
- **train_test_split**: From sklearn.model_selection, used to split data into training and testing sets.
- **Various Classifiers**: Imported from sklearn.tree, sklearn.svm, sklearn.linear_model, sklearn.ensemble, sklearn.neighbors, and sklearn.naive_bayes. These are different algorithms used for classification.
- **accuracy_score**: From sklearn.metrics, used to calculate the accuracy of the models.

3. Loading the Dataset

```
# Load the dataset from a CSV file
df = pd.read_csv('mobile_price_range_data.csv')
```

- Reads the dataset mobile_price_range_data.csv into a pandas DataFrame (df).

4. Separating Features and Target

```
# Separate features and target
X = df.drop('price_range', axis=1) # Features
y = df['price_range'] # Target variable
```

- **X**: Contains all columns except price_range, which are the features.
- **y**: Contains only the price_range column, which is the target variable (the class labels).

5. Splitting the Dataset into Training and Testing Sets

```
# Split the dataset into training and testing sets
X_train, X_test, y_train, y_test = train_test_split(X, y, test_size=0.3, random_state=42)
```

- **train_test_split**: Splits the dataset into training (70 per cent) and testing (30 per cent) sets.
- test_size=0.3 indicates 30 per cent of the data will be used for testing.
- random_state=42 ensures reproducibility by fixing the random seed.

6. Initializing and Training the Classifiers

```
# Initialize the classifiers
classifiers = {
    "Decision Tree": DecisionTreeClassifier(),
    "SVM": SVC(),
    "LR Multinomial": LogisticRegression(multi_class='multinomial', solver='lbfgs', max_iter=200),
    "Random Forest": RandomForestClassifier(),
    "K-Nearest Neighbors": KNeighborsClassifier(),
    "Naive Bayes": GaussianNB()
}
# Train the classifiers
for name, clf in classifiers.items():
    clf.fit(X_train, y_train)
```

- **classifiers**: A dictionary where keys are classifier names and values are instances of respective classifiers initialized with default parameters.
- **Training**: Each classifier (clf) is trained on the training set (X_train, y_train).

7. Evaluating the Classifiers

```
# Evaluate the classifiers
accuracies = {name: accuracy_score(y_test, clf.predict(X_test)) for name, clf in classifiers.items()}
```

- **accuracy_score**: Calculates the accuracy of predictions compared to actual y_test values.
- **accuracies**: Dictionary where keys are classifier names and values are accuracy scores on the test set (X_test, y_test).

8. Function for User Input

```
# Function to take user input for features
def get_user_input():
    features = [
        "battery_power", "clock_speed", "dual_sim", "four_g", "int_memory",
        "mobile_wt", "n_cores", "px_height", "px_width", "ram", "three_g",
        "touch_screen", "wifi"
```

```
    ]
    input_data = []
    print("Please enter the following features: ")
    for feature in features:
      value = float(input(f"{feature}: "))
      input_data.append(value)
    return [input_data]
```

- **get_user_input()**: Prompts the user to input values for mobile phone features defined in the features list.

9. Making Predictions with User Input

```
# Get user input
user_input = get_user_input()

# Make predictions with user input
predictions = {name: clf.predict(user_input)[0] for name, clf in classifiers.
items()}
```

- **user_input**: Contains user-provided feature values for a new mobile phone.
- **predictions**: Dictionary where keys are classifier names and values are predicted price_range based on user_input.

10. Displaying Results

```
# Display the predictions and accuracies
print("\nClassifier Accuracies: ")
for name, accuracy in accuracies.items():
    print(f"{name} Accuracy: {accuracy: .2f}")

print("\nPredictions based on your input: ")
for name, prediction in predictions.items():
    print(f"{name} Prediction: {prediction}")
```

- **Output**: Displays accuracies of classifiers on the test set and predictions based on user input.

Example Outputs

Classifier Accuracies

- ☆ Classifier Accuracies:
- ☆ Decision Tree Accuracy: 0.81
- ☆ SVM Accuracy: 0.95
- ☆ LR Multinomial Accuracy: 0.62
- ☆ Random Forest Accuracy: 0.90
- ☆ K-Nearest Neighbors Accuracy: 0.92
- ☆ Naive Bayes Accuracy: 0.79

Accuracy scores for each classifier evaluated on the test set (X_test, y_test).

Predictions Based on User Input

Please enter the following features:

battery_power: 1500

clock_speed: 2.0

dual_sim: 1

four_g: 1

int_memory: 16

mobile_wt: 150

n_cores: 4

px_height: 800

px_width: 1200

ram: 2048

three_g: 1

touch_screen: 1

wifi: 1

- ☆ Predictions based on your input:
- ☆ Decision Tree Prediction: 2
- ☆ SVM Prediction: 2
- ☆ LR Multinomial Prediction: 2
- ☆ Random Forest Prediction: 1
- ☆ K-Nearest Neighbors Prediction: 2
- ☆ Naive Bayes Prediction: 1

Predicted price_range for a new mobile phone based on user-provided feature values using each classifier.

Data Description

battery_power: total energy a battery can store in one time measured in mAh

clock_speed: speed at which microprocessor executes instructions

dual_sim: support dual sim or not

four_g: support 4G or not

int_memory: Internal Memory in Gigabytes

mobile_wt: weight of mobile phone

n_cores: number of core processor

px_weight: pixel resolution height

px_width: pixel resolution width

ram: random access memory in megabytes

three_g: support 3G or not

touch_screen: has touch screen or not

wifi: has wifi or not

price_range: this is the target variable with value of 0(low cost), 1(medium cost), 2(high cost) and 3(very high cost).

Summary

This example demonstrates a machine learning workflow for classifying mobile phone price ranges.

Data Preprocessing

1. Loads a dataset from a CSV file.
2. Separates features and the target variable.
3. Splits the dataset into training and testing sets.

Model Training

1. Initializes various classification models (Decision Tree, SVM, Logistic Regression, Random Forest, Gradient Boosting, K-Nearest Neighbors, Naive Bayes).
2. Trains these models on the training data.

Model Evaluation

1. Evaluates the models using accuracy scores on the testing data.

User Input

1. Collects feature values from the user for a new mobile phone.

Prediction

1. Uses the trained models to predict the price range of the new mobile phone based on the user's input.
2. Displays the predictions and the accuracy of each model.

Chapter 14

Object Detection Using Roboflow and YOLOv8

Object Detection

Object detection is a key field in Artificial Intelligence, allowing computer systems to "see" their environments by detecting objects in visual images or videos. Object detection is often called object recognition, object identification, and image detection, and these concepts are synonymous.

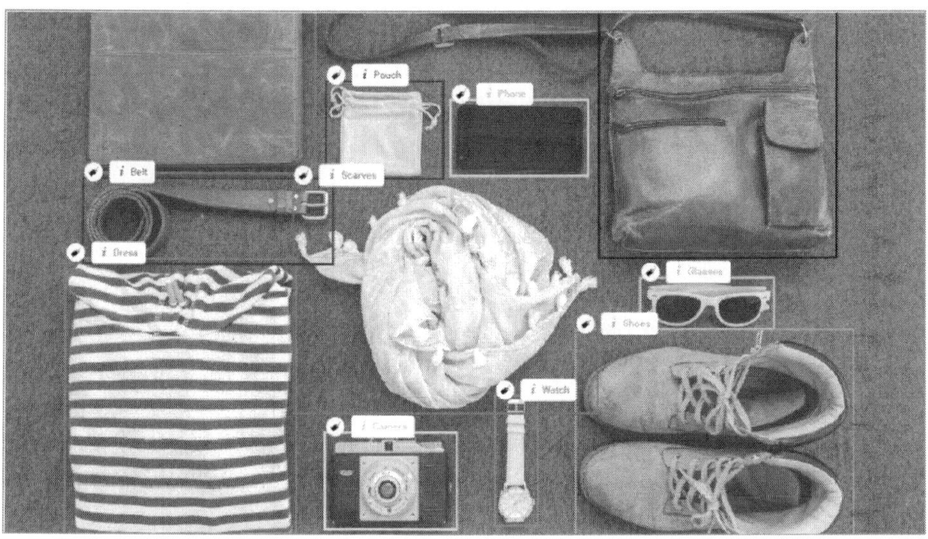

Humans can easily detect and identify objects present in an image. The human visual system is fast and accurate and can perform complex tasks like identifying multiple objects and detecting obstacles with little conscious thought. With the availability of large amounts of data, faster GPUs, and better algorithms, we can quickly train computers to detect and classify multiple objects within an image with high accuracy.

With this identification and localisation, object detection can count objects in a scene and determine and track their precise locations while accurately labelling them.

Object detection is a key field in Artificial Intelligence, allowing computer systems to "see" their environments by detecting objects in visual images or videos. Object detection is often called object recognition, object identification, and image detection; these concepts are synonymous.

What is Object Detection?

Object detection is an important computer vision task used to detect instances of visual objects of certain classes (for example, humans, animals, cars, or buildings) in digital images such as photos or video frames. Object detection aims to develop computational models that provide the essential information needed by computer vision applications: "What objects are where?".

Object detection is not, however, akin to other common computer vision technologies such as classification (assigning a single class to an image), keypoint detection (identifying points of interest in an image), or semantic segmentation (separating the image into regions via masks).

As with every emerging tech, plenty of technical terms might confuse or be thought of as synonyms for computer vision. There's classification, detection, tracking, counting, and more. However, object detection and image classification are the biggest confusion points. At the most basic level, the difference between classification and detection is simple:

Image classification: Applies a prediction to an image based on an analysis of the contents.

Objection detection: Locates objects within an image.

Why is Object Detection Important?

Object detection is one of the fundamental problems of computer vision. Moreover, it forms the basis of many other downstream computer vision tasks, for instance, segmentation, image captioning, object tracking, *etc.* Specific object detection applications include pedestrian detection, people counting, face detection, text detection, pose detection, and number-plate recognition.

How are object recognition models trained?

The AI model training process for object recognition is similar to that of image recognition. However, one crucial difference is the labels for the input dataset.

Object recognition datasets bundle together an image or video with a list of objects it contains and their locations.

Before training an object recognition model, machine learning experts must decide which categories they would like the AI model to recognise. For example, a simple mask detection model might classify faces in images as "with mask," " or "without a mask." Each faces in the image or video in the training dataset needs to be associated with one of these labels so the model can learn it during the training process.

Once the object recognition model is trained, it can analyse real-world data. For example, the model accepts an image as input and returns a list of predictions for the image's label. The more data you give your model, the better your device will be at recognising the objects you want and learning how to improve for the future.

Object Detection Use Cases and Applications

The use cases involving object detection are very diverse. There are almost unlimited ways to make computers see like humans to automate manual tasks or create new, AI-powered products and services. It has been implemented in computer vision programs for various applications, from sports production to productivity analytics. Object recognition is the core of most vision-based AI software and programs today. Object detection plays a vital role in scene understanding, popular in security, transportation, medical, and military use cases.

- ☆ Object detection in retail: Strategically placed people counting systems throughout multiple retail stores are used to gather information about how customers spend their time and customer footfall. AI-based customer analysis to detect and track customers with cameras helps to understand customer interaction and experience, optimise the store layout, and make operations more efficient. A widespread use case is the detection of queues to reduce waiting time in retail stores.
- ☆ Autonomous driving: Self-driving cars depend on object detection to recognise pedestrians, traffic signs, other vehicles, and more. For example, Tesla's Autopilot AI heavily utilises object detection to perceive environmental and surrounding threats, such as oncoming vehicles or obstacles.
- ☆ Video surveillance: Because state-of-the-art object detection techniques can accurately identify and track multiple instances of a given object in a scene, these techniques naturally lend themselves to automated video surveillance systems. For example, object detection models can track multiple people at once, in real-time, as they move through a given scene or across video frames. From retail stores to industrial factory floors, this

kind of granular tracking could provide invaluable insights into security, worker performance and safety, retail foot traffic, and more. Example of object detection in video analytics for people detection in dangerous areas using CCTV cameras

- ☆ Vehicle detection with AI in transportation: Object recognition is used to detect and count vehicles for traffic analysis or to detect cars that stop in dangerous areas, such as crossroads or highways.
- ☆ Animal detection in agriculture: Object detection is used in agriculture for tasks such as counting, animal monitoring, and evaluation of the quality of agricultural products. Damaged produce can be detected while processing using machine learning algorithms.
- ☆ Medical feature detection in healthcare: Object detection has allowed for many breakthroughs in the medical community. Because medical diagnostics rely heavily on studying images, scans, and photographs, object detection involving CT and MRI scans has become extremely useful for diagnosing diseases, for example, with ML algorithms for tumour detection.

Object detection in YOLO-v8 model

You Only Look Once is an acronym that stands for YOLO. Version 8 used, which, as of this writing, is the most sophisticated object identification algorithm available, was introduced by Ultralytics inJune 2020. It is a brand-new object detection convolutional neural network (CNN). This method evaluates the entire image using a single neural network, thenbreaks it downinto its component parts and forecasts bounding boxes and probabilities for eachone independently. The expected probability is used to weight these bounding boxes. The approach "only looks once" at the image inthe sense that predictions are made after just one forward propagationrunthroughthe neural network. After non-max uppression, it then delivers detected items (which ensures that the object detection algorithm only identifies each object once). YOLO is a cutting-edge, real-time object detector, and YOLOv8 is built onYOLOv1–YOLOv7. It has achieved top performance ontwo official object detectiondatasets: Pascal VOC (visual object classes) (Everingham *et al.*, 2015) and Microsoft COCO (common objects in context) (Lin *et al.*, 2014). Incontrast to other cutting-edge detectiontechniques, like Fast R-CNN (Region Based Convolutional Neural Network), YOLO considers the entire image rather than concentrating on just one region of interest, which results in fewer background errors.

YOLOv8 -CNN Architecture

The YOLOv8 architecture expands onthe YOLO algorithms' earlier iterations. The convolutional neural network used by YOLOv8 is composed of two primary sections: the head and the backbone. The foundationof YOLOv8 is a modified version of the CSPDarknet53 architecture and 53 convolutional layers make up this design, whichalso uses cross-stage partial connections to enhance information transfer between the various layers. A few convolutional layers and a string of fully connected layers make up the head of the YOLOv8 algorithm. For the objects that are discovered inanimage, these layers oversee predicting bounding boxes, objectness scores, and class probabilities. The usage of a self-attention mechanism inthe network's brainis one of YOLOv8's distinguishing characteristics. Withthe aid of this mechanism, the model canchange its attentionto various aspects of the image according to their relevance to the task at hand. The capability of YOLOv8 to carry out multi-scale object identificationis another crucial feature. The model makes use of a feature pyramid network to find items inanimage that are of different sizes and scales. Figure represents the YOLOv architecture. The numerous layers of this feature pyramid network detect items at various scales, enabling the model to recognize bothlarge and small things in an image. Object typically range insize from little to medium. Due to multi-scale detection, the model candetect size differences during the Object detecting process.

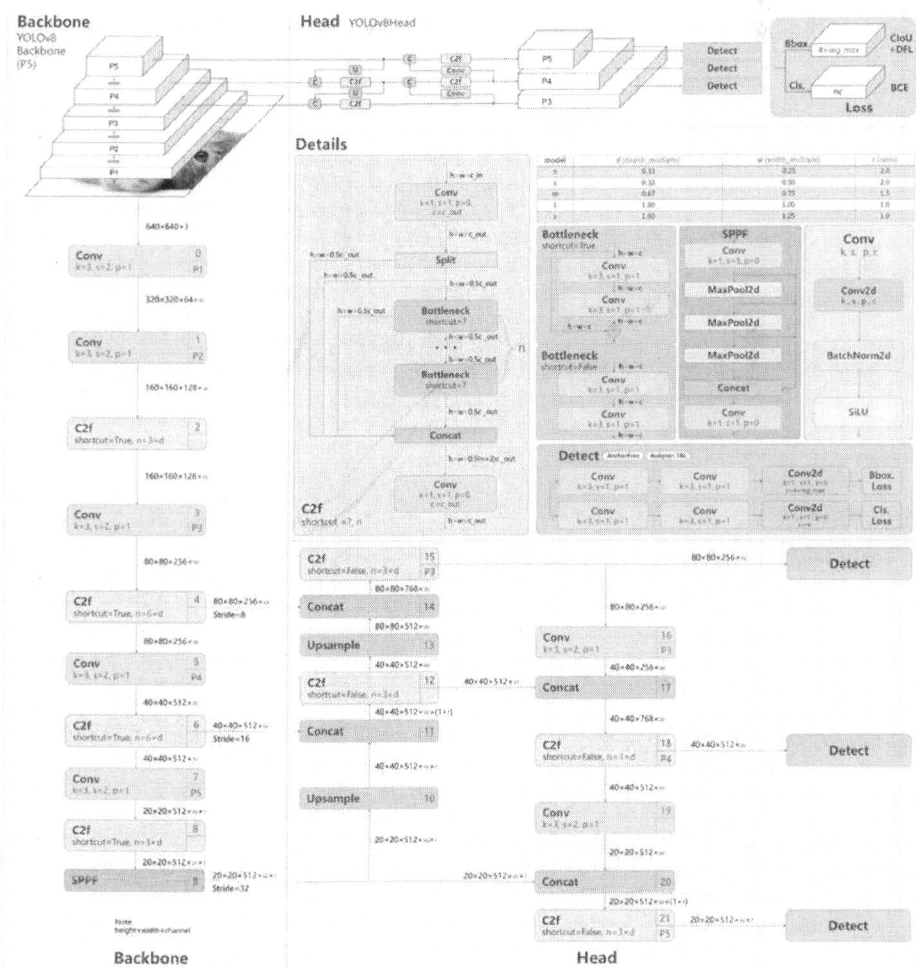

YOLOv8 Architecture

Depicts the three-step process used by YOLO to recognize and classify things. The image is initially divided into an S X S sized grid. The number of bounding boxes in each cell is predicted using the variable B once the image has been divided into a grid. In grid cells that have totally contained items, object detection takes place. The YOLOv8-Seg model is a development of the YOLOv8 object identification model that additionally segments the input image in a semantic manner.The old YOLO neck architecture is replaced by a unique C2f module as the YOLOv8-Seg model's backbone, which is a CSPDarknet53 feature extractor. Two segmentation heads that learn to anticipate the semantic segmentation masks for the input image come after the C2f module. The model has five detection modules and a prediction layer, which are identical to the YOLOv8 detection heads. On numerous object detection and semantic segmentation benchmarks, it has been demonstrated that the YOLOv8-Seg model achieves state-of-the-art performance while maintaining high speed and economy. The results of the detection (class, score, position, and size) are then output by YOLO Layer.

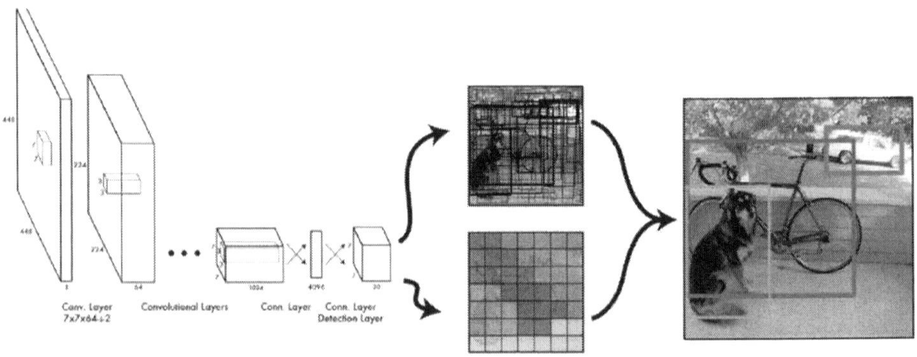

YOLO Three Step Object Detection Process

Roboflow

Roboflow is a Computer Vision developer framework for better data collection to preprocessing, and model training techniques. Roboflow has public datasets readily available to users and has access for users to upload their own custom data also. Roboflow accepts various annotation formats. In data pre-processing, there are steps involved such as image orientations, resizing, contrasting, and data augmentations. The entire workflow can be co-ordinated with teams within the framework. For model training, there's a bunch of model libraries already present such as EfficientNet, MobileNet, Yolo, TensorFlow, PyTorch, *etc.* Thereafter model deployment and visualization options are also available hence encompassing the entire state-of-art. Roboflow is used in various computer vision industries for use cases such as – gas leak detection, plant vs weed detection, aeroplane maintenance, roof damage estimator, satellite imagery, self-driving cars, traffic counter, garbage cleaning, and many more. Steps To Use Roboflow in Object Detection: 1.Dataset Loading 2.Labeling 3.Organise 4.Process 5.Train

Roboflow provides two services that can help you collect data for YOLOv8 models: Universe and Collect.

Universe is an online repository with over 250,000 vision datasets totalling over 100 million images.

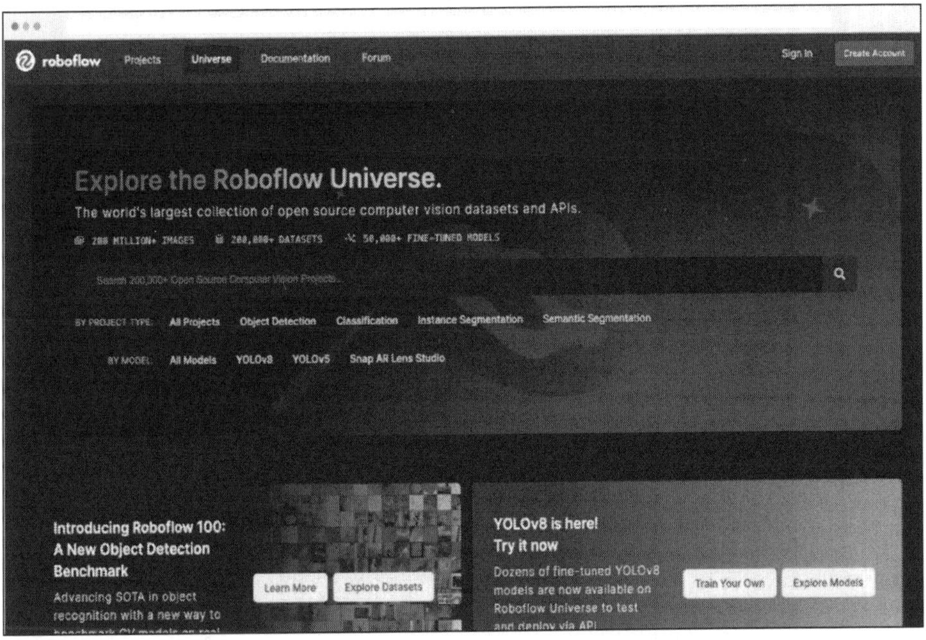

With a free Roboflow account, you can export any dataset available on Universe. To export a dataset, click the "Download this Dataset" button on any dataset.

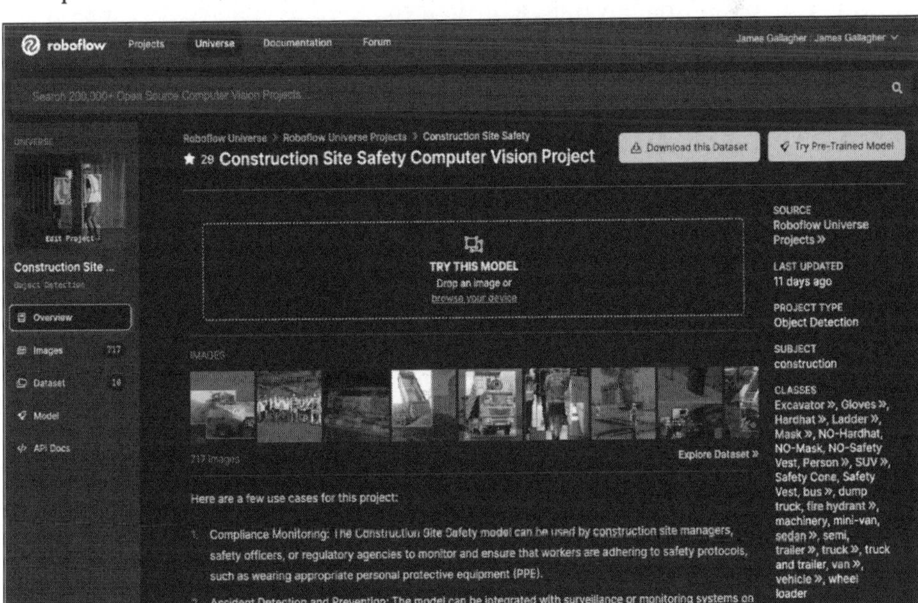

For YOLOv8, select "YOLOv8" as the export format:

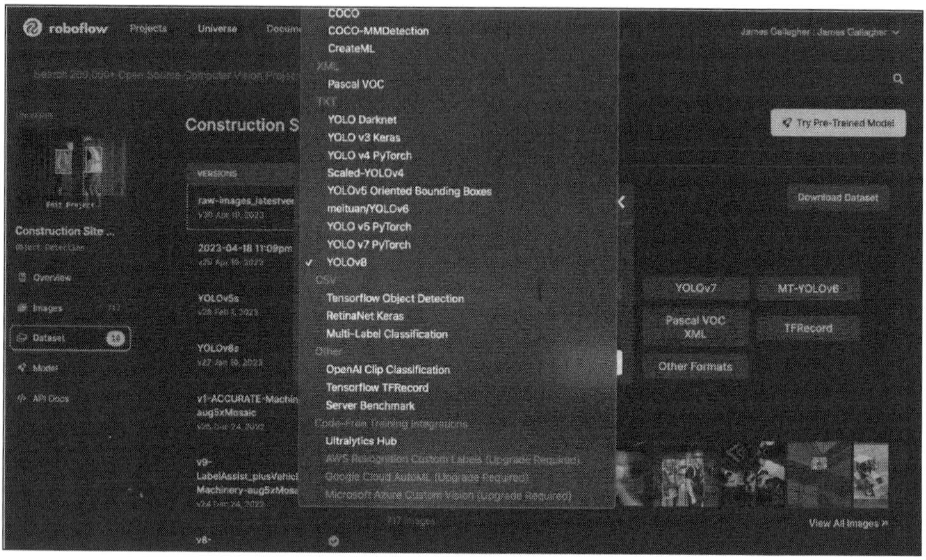

Universe also has a page that aggregates all public fine-tuned YOLOv8 models uploaded to Roboflow. You can use this page to explore pre-trained models you can use for testing or for automated data labeling or to prototype with Roboflow inference.

If you want to gather images yourself, try Collect, an open source project that allows you to automatically gather images using a webcam on the edge. You can use text or image prompts with Collect to instruct what data should be collected, allowing you to capture only the useful data you need to build your vision model.

Upload, Convert and Label Data for YOLOv8 Format

Roboflow Annotate is an online annotation tool for use in labeling images for object detection, classification, and segmentation.

To label data for a YOLOv8 object detection, instance segmentation, or classification model, first create a project in Roboflow.

Next, upload your images, and any pre-existing annotations you have from other tools (using one of the 40+ supported import formats), into Roboflow.

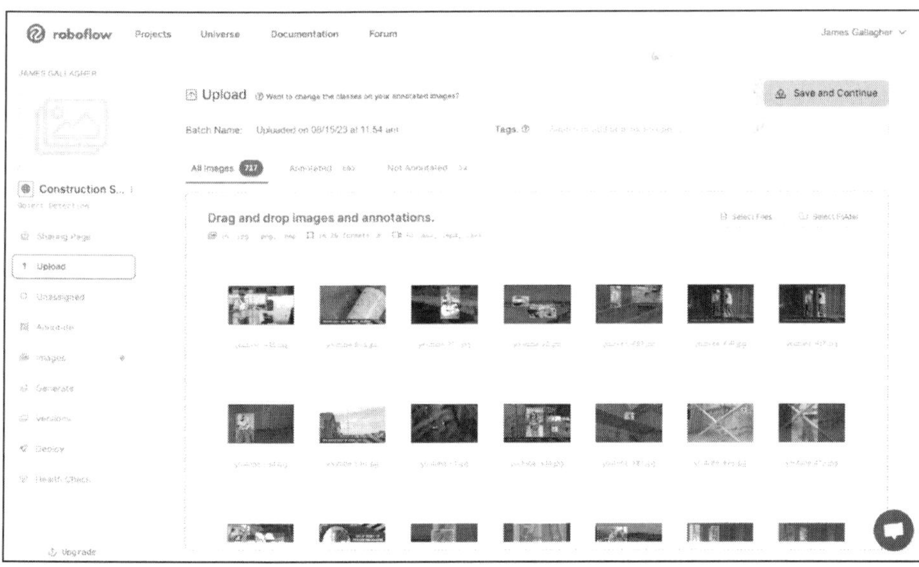

Select the batch of images you have uploaded on the Annotate page to which you are taken after uploading images. Then, click "Start Annotating" to label images.

To label with bounding boxes, press the B key on your keyboard or click the box icon in the sidebar. Click on a point where you want to start your bounding box, then drag to create the box:

A pop-up will appear asking you to select a class for your annotation once you have created an annotation.

To label with polygons, press the P key on your keyboard, or the polygon icon in the sidebar. With the polygon annotation tool enabled, click on individual points in the image to draw a polygon.

Roboflow offers a SAM-based label assistant with which you can label images faster than ever. SAM (Segment Anything Model) is a state-of-the-art computer vision model that can precisely label images. With SAM, you can significantly speed up the image labeling process. Annotating images with polygons becomes as simple as a few clicks, rather than the tedious process of precisely clicking points around an object.

To use the label assistant, click the cursor icon in the sidebar, SAM will be loaded for use in your project.

Hover over any object in the image and SAM will recommend an annotation. You can hover to find the right place to annotate, then click to create your annotation. To amend your annotation to be more or less specific, you can click inside or outside the annotation SAM has created on the document.

You can also add tags to images from the Tags panel in the sidebar. You can apply tags to data from a particular area, taken from a specific camera, and more. You can then use these tags to search through data for images matching a tag and generate versions of a dataset with images that contain a particular tag or set of tags.

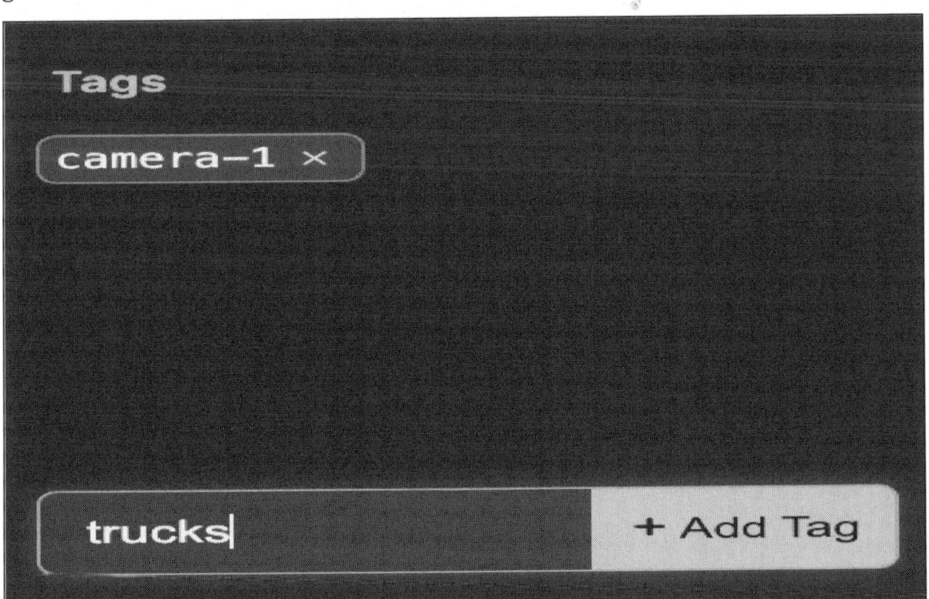

Models hosted on Roboflow can be used with Label Assist, an automated annotation tool that uses your YOLOv8 model to recommend annotations. To use Label Assist, first upload a YOLOv8 model to Roboflow (see instructions later in the guide). Then, click the magic wand icon in the left sidebar and select your model for use in Label Assist.

Choose a model, then click "Continue" to enable Label Assist:

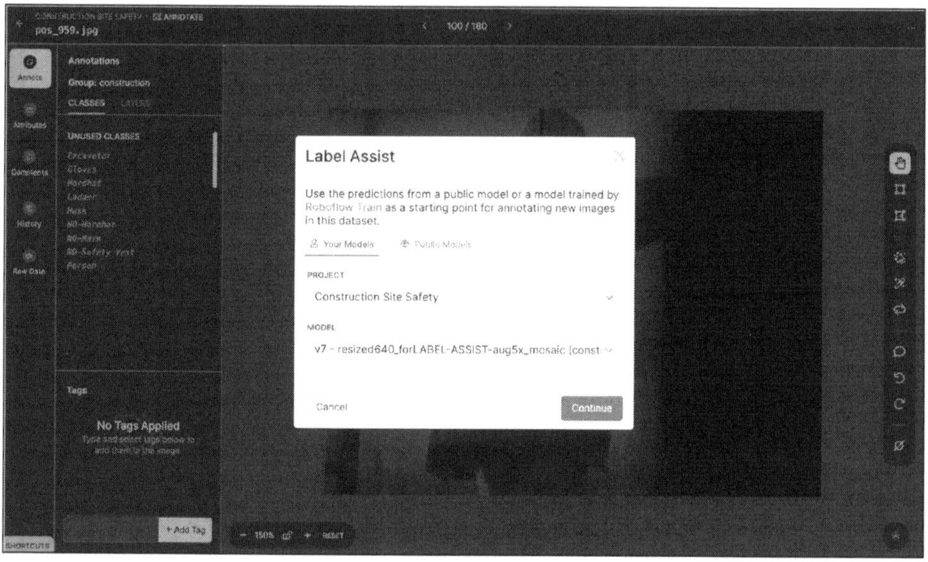

When you open new images for annotation, Label Assist will trigger and recommend annotations.

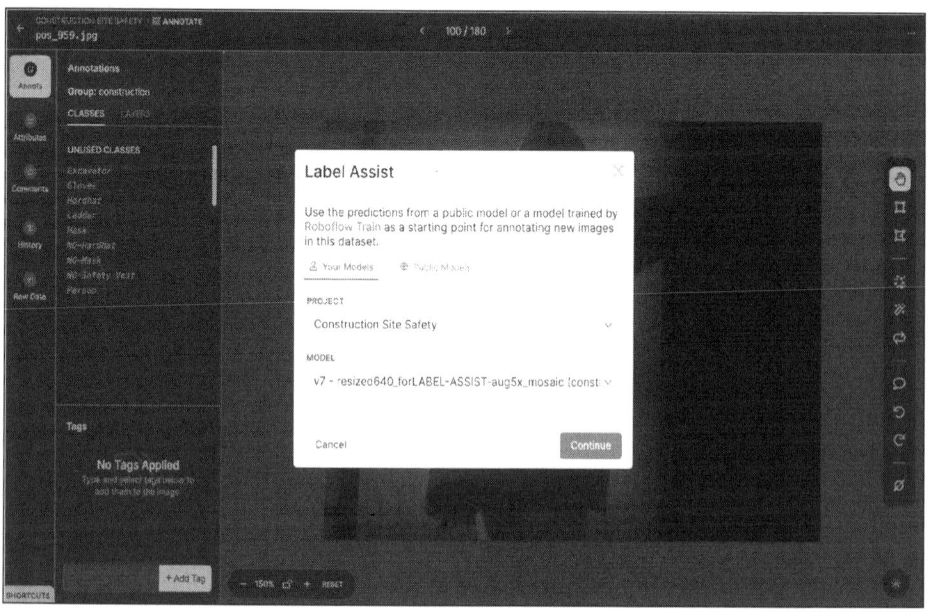

Dataset Management for YOLOv8

Roboflow provides a suite of tools for understanding computer vision datasets.

First, you can use dataset search to find images that meet a semantic text description (*i.e.* find all images that contain people), or that meet a specified label (*i.e.* the image is associated with a specific tag). To use dataset search, click "Dataset" in the sidebar. Then, input a search query using the search bar and associated filters at the top of the page.

For example, the following text query finds images that contain people in a dataset:

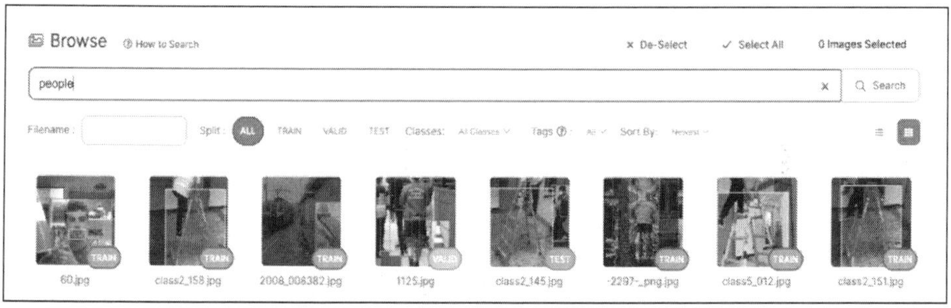

You can narrow your search to images with a particular tag using the "Tags" selector:

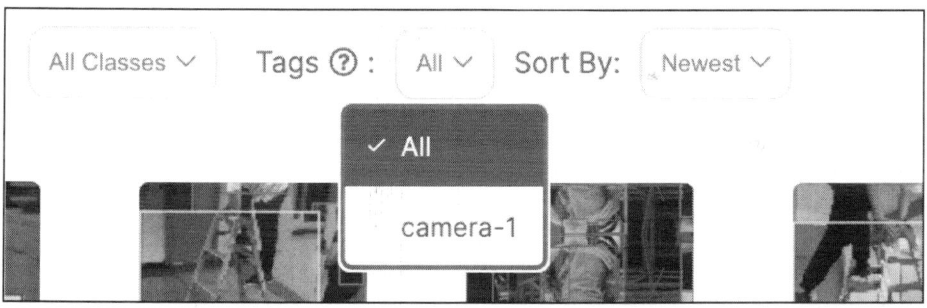

Before you start training a model with your dataset, we recommend using Roboflow Health Check, a web tool that provides an insight into your dataset and how you can improve the dataset prior to training a vision model.

To use Health Check, click the "Health Check" sidebar link. A list of statistics will appear that show the average size of images in your dataset, class balance, a heatmap of where annotations are in your images, and more.

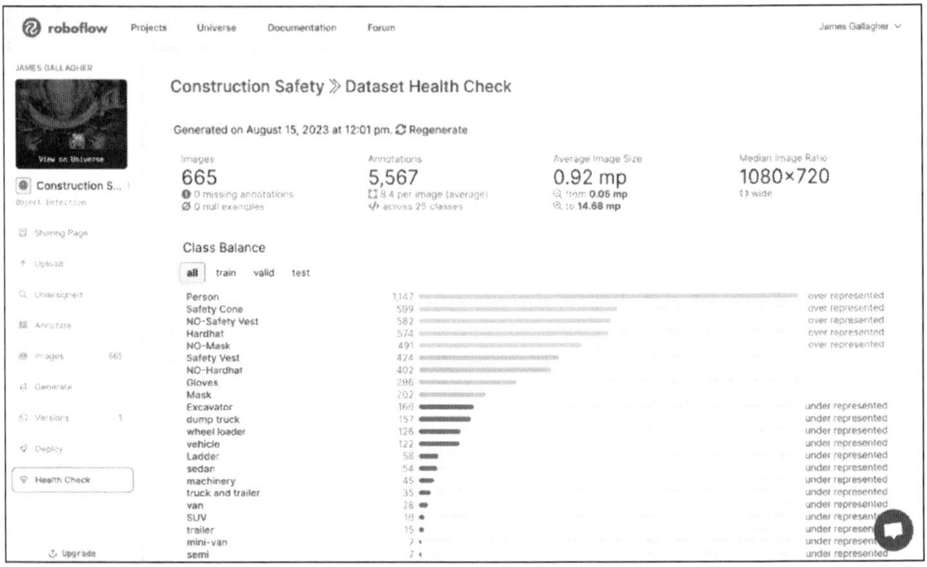

Health Check may recommend changes to help enhance dataset performance. For example, the class balance feature may show that there is an imbalance in labels that, if solved, may boost performance or your model.

Export Data in 40+ Formats for Model Training

To export your data, you will need a dataset version. A version is a state of your dataset frozen-in-time. To create a version, first click "Versions" in the sidebar. Then, click the "Create New Version" button. On this page, you will be able to choose augmentations and preprocessing steps to apply to your dataset:

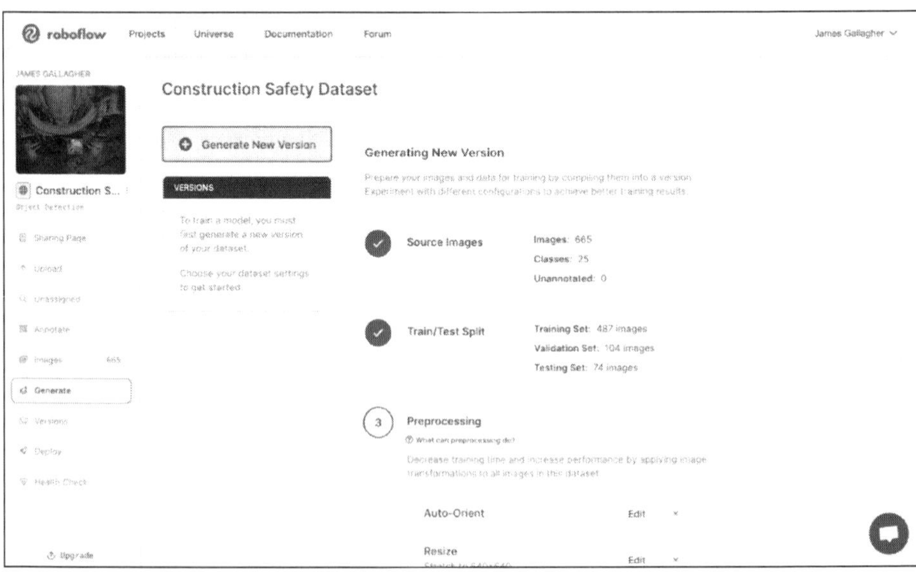

For each augmentation you select, a pop-up will appear allowing you to tune the augmentation to your needs. Here is an example of tuning a brightness augmentation within specified parameters:

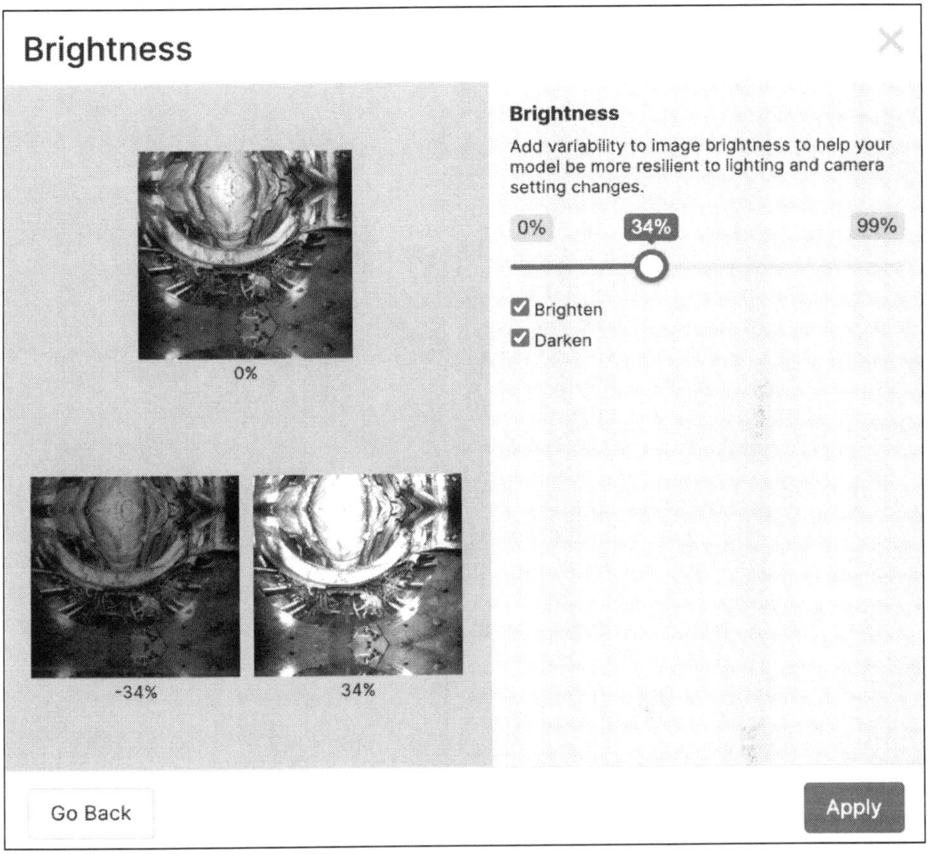

When your dataset version has been generated, you can export your data into a range of formats. Click the "Export Dataset" button on your dataset version page to export your data:

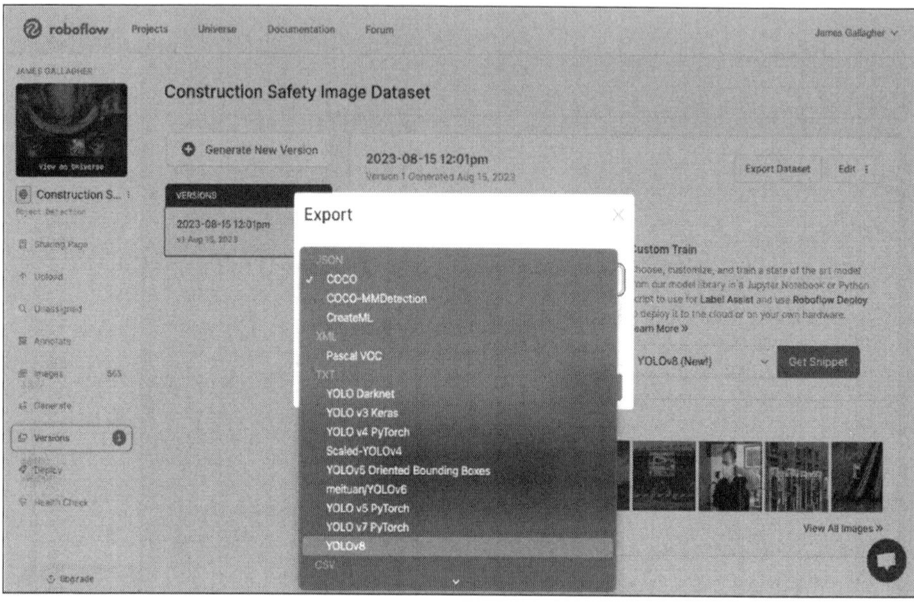

You are now ready to train YOLOv8 on a custom dataset.

Data Handling

Dataset Creation

To prepare this unique image dataset for this study, about 1704 images/video were framed of Object from the web. For this ground truth bounding box, software roboflow used.

Roboflow (Free, Open-Source Tool for Graphically Labelling Images). It is written in python as its graphical interface. The main work done by Roboflow for this study is to label/prepare/host/automatically YOLO formatted custom data set (Object) for this study.

Train with more than 5000 instances for each object class and 1000 images for each class to create a robust YOLOov8 model.

YOLO Labelling Format

The majority of annotation platforms allow for the export of one text file with annotations per image in the YOLO labelling format. For each object in the image, a bounding-box (BBox) annotation is present in a separate text file. The annotations, which range from 0 to 1, are normalised to the size of the image. They are represented in the following format: (Object class ID) (X-center) (Y-center)

(Boxwidth) (Box-height). If there is only one object in the image (Object) the YOLO annotations text file's.

Object-class-ID	X-Center	Y-Center	Box-width	Box-height
1	0.315263	0.603174	0.018346	0.057229
1	0.326109	0.514789	0.010091	0.042734
1	0.354533	0.145287	0.016343	0.036621
1	0.422650	0.785424	0.032643	0.078637
1	0.57837	0.437988	0.059572	0.041039

The unit of the normalized annotations is a percentage of the image size. The X-center and Y-center coordinates are the center of the bounding box relative to the top-left corner of the image, expressed as a percentage of the image width and height, respectively. The Box-width and Box-height are the width and height of the bounding box relative to the image width and height, respectively, also expressed as a percentage.

Data Directory's Structure

To comply with Ultralytics directory's structure, the data is provided in the following structure.

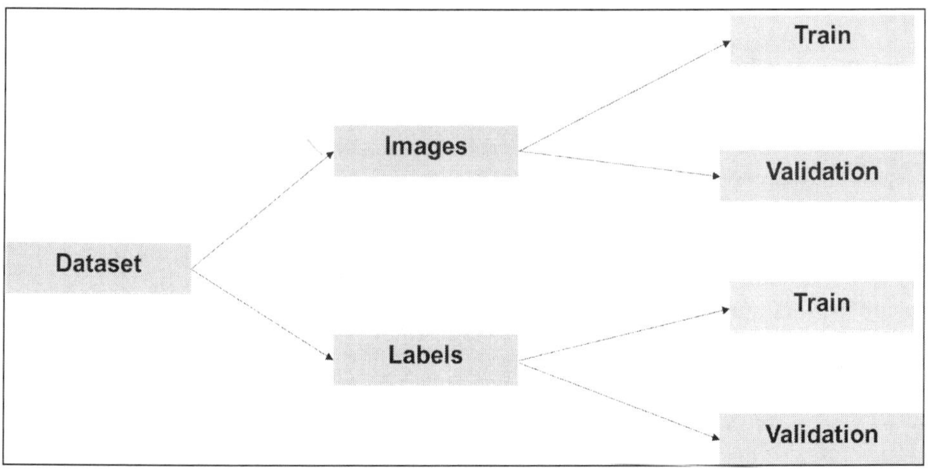

Training

For this study, the YOLOv5s model was used, which is a commonly used technique. Training YOLOv5 on a custom dataset is a relatively easy and straightforward process. Here are the steps involved:

Training from Scratch

The model will perform best when it is trained entirely from scratch with a sufficiently large dataset. By giving the weights argument an empty string, the weights are initialized at random.

- ☆ Batch: batch size
- ☆ Epoch: number of epochs.
- ☆ Data: path to the data-configurations file.
- ☆ cfg: path to the model-configurations file.
- ☆ Weights: path to initial weights.
- ☆ Cache: cache images for faster training.
- ☆ Img: image size in pixels.

Training from a Pre-Trained Model

Although there is support for additional pre-trained models like VOC (Visual Object Classes), COCO (Common Objects in Context) is a dataset for object detection that includes pictures of commonplace settings. It has 123 classes, including animals like cats, dogs, elephants, and zebras. However, there is no class for Object. By giving the name of the pre-trained COCO model to the "weights" option, the model was initialised with weights from that model. The pre-trained model will be downloaded automatically.

Feature Extraction

The backbone layer, which acts as a feature extractor, and the head layer, which computes the output predictions, make up the two fundamental components of a model. The same backbone as the pre-trained COCO model will be utilized, and the model's head will be solely trained to further offset the tiny dataset size. The "freeze" parameter will fix the 12 layers that make up the YOLOv8s backbone.

Validation

The validation script will be used to rate the model. Depending on the 'task' input, performances can be assessed across the training, validation, or test dataset Splits.

Dataset Distribution

Dataset	*Augmentation*	*Images*
Training	No	1391
Validation	No	275
Test	No	72

It represents the splitting of different datasets based on augmentation or not augmentation. In order to apply object detection in this study, 1738 images from internet were collected. Then the images were divided into training (80 per cent) and validation (16 per cent) and test (4 per cent) dataset. In order to train the model's upper layers to recognise the classes, the rest area dataset were added and the number of training epochs were modified and stack size. Split of the data into training, validation and test dataset model was trained for 50 epochs, and it took about 40 minutes to complete the run.

Object Detection by Deep Learning

The integration of machine learning and deep learning has significantly improved farm surveillance. Object detection algorithms have proven invaluable in this regard, as they can accurately track and count Object even in challenging underwater conditions such as turbid water, darkness, and nighttime. One such advanced algorithm that has been employed is YOLOv8, which relies on Convolutional Neural Networks (CNN). Its primary function is to detect objects based on their texture and boundary edges, irrespective of their movement. The model was trained using Google Colab, which provides access to powerful GPUs at no cost and requires no additional configuration. The study utilized a Roboflow.ai notebook based on YOLOv8 and incorporated pre-trained COCO (Common Objects in Context) weights. The YOLO algorithm divides the image into a grid of equal-sized grids, with each grid responsible for detecting and locating objects within it. In contrast to Faster R CNN and similar algorithms that use the Region Proposal Network to identify potential regions of interest before recognition, YOLO makes all predictions through a single fully connected layer. In the realm of object detection, various sensors and cameras are readily available to identify any type of motion. However, this motion can be triggered not only by Object but also by other objects like leaves or logs passing through the frame. And then the weights of detection model used for tracking and counting in the same model. Consequently, there is a risk of either overestimating the Object quantity or arbitrarily excluding some registered objects as debris, potentially leading to inaccurate results. Another challenge arises from the fact that Object can swim rapidly and have a whitish coloration. To address these issues, the

YOLOv8 algorithm, which is grounded in Convolutional Neural Networks (CNN), was employed. This algorithm specializes in object detection based on texture and boundary edges, independent of their motion. Data set divided into training, validation, test dataset and preprocessing and augmentation was done for training dataset

Split of the Data into Training, Validation and Test Dataset

Dataset	*Augmentation*	*Images*
Training	No	1391
Validation	No	275
Test	No	72

The model was trained for 50 epochs, and it took about 40 minutes to complete the run, showing the graphs, which show various performance measures for the training set, validation data set, shows how the box loss and objectness loss is reducing.

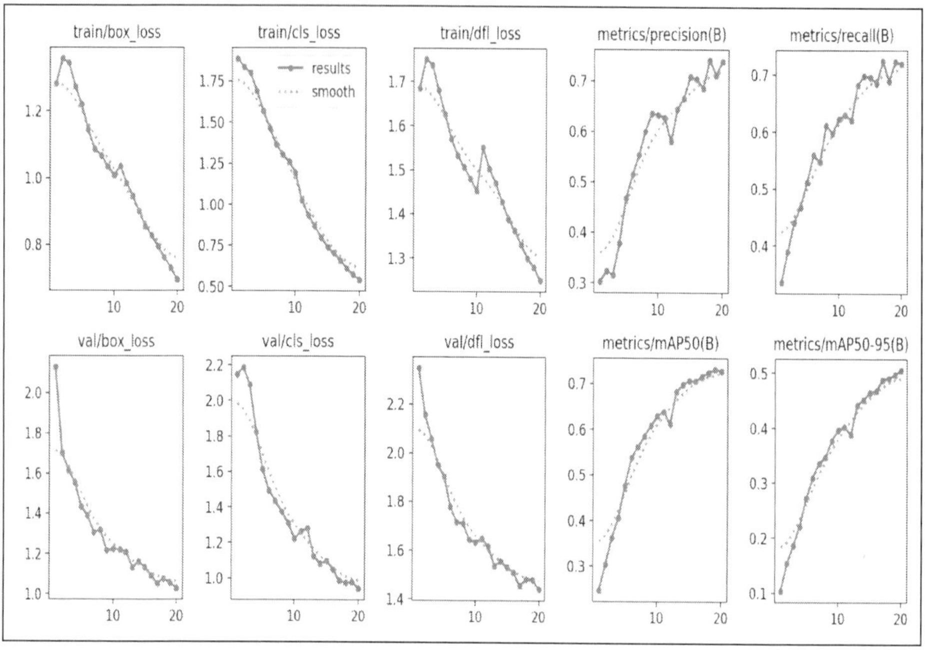

Plots of Box Loss, Objectness Loss, Classification Loss, and Mean Average Precision (mAP) over the Training Epochs for the Validation Dataset.

Displays three basic types of loss: classification loss, objectness loss, and box loss for both training data and validation data. The box loss measures how accurately the algorithm can pinpoint an object's center and how completely the predicted bounding box encloses an object. Objectness essentially measures the likelihood that an object exists in each zone of interest. If the objectivity is high, an item is probably present in the image window. How successfully the algorithm can determine the proper class of a given object is shown by the classification loss. Objectness loss measures the probability that an object exists in a proposed region of interest. A low objectness loss indicates that the model is confident that an object exists in the region. Mean precision measures (Mpi) the average precision (APi) of the model's predictions. A high mean precision indicates that the model is accurate in its predictions. Box loss measures the difference between the predicted bounding boxes and the ground truth bounding boxes. A low box loss indicates that the model's predicted bounding boxes are close to the ground truth bounding boxes.

Precision measures the percentage of positive predictions made by the model that are correct and recall measures the percentage of all actual positive instances that were correctly identified by the model.

Model Evaluation Predictors

$$mAP\ (mean\ average\ precision) = \frac{1}{N}\sum_{i=1}^{N} AP_i$$

$$F_1\ score = 2 \times \frac{(Precision \times recall)}{(Precision + recall)}$$

$$Precision = \frac{(True\ positive)}{(True\ positive + false\ positive)}$$

$$Recall = \frac{(True\ positive)}{(True\ positive + false\ negative)}$$

This values show of objectness loss is slightly higher than the train objectness loss. This indicates that the model is not as confident in its predictions on the validation set as it is on the training set. The value of mean precision is slightly lower than the train mean precision. This indicates that the model is not as accurate in its predictions on the validation set as it is on the training set. The value of box loss is slightly higher than the train box loss. This indicates that the model's predicted bounding boxes are not as close to the ground truth bounding boxes on the validation set as they are on the training set. The proportion of accurately categorized positive samples (True Positive) to the total number of positively classified samples (either correctly or incorrectly) is known as precision. Precision is a crucial metric in evaluating model performance, and a higher precision score indicates a better ability to classify positive samples accurately. On the other hand, recall is determined as the proportion of positive samples that were correctly identified as positive to all positive samples. Recall estimates how well the model can identify positive samples, emphasizing the importance of not missing any positive cases. A higher recall in train dataset than validation dataset indicates that the model can successfully identify slightly less positive samples accurately in validation than train data set. But the recall value is quite high for both which indicates the good fit model with F score of 0.849 Overall, the results of the training and validation curves are promising. The model is learning to predict objectness, precision, and bounding boxes more accurately as the number of epochs increases. However, there is still some room for improvement, especially on the validation set. After the model was trained, testing was conducted using new and unseen test images and videos. The results can be seen below.

Identified Objects on Test Data Set with Accuracy

The model's performance for tracking and counting Object behavior and their trajectories in stress condition in real-time was evaluated using a video recorded by the experimental setup camera. The model efficiently recognized Object in each frame with a high degree of accuracy.

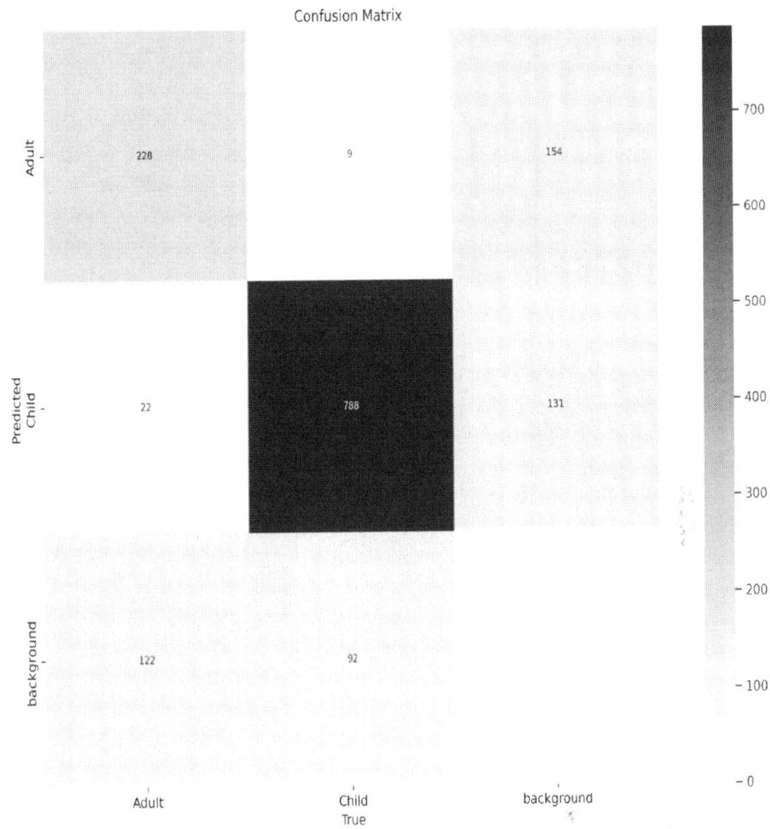

Confusion Matrix for Object Detection Model

A confusion matrix is a table that is used to evaluate the performance of a machine learning classification model. It shows the number of correct and incorrect predictions made by the model, broken down by each class.

- True positive (TP): This is an example of detection for which the classifier correctly predicted a positive outcome when a positive outcome really occurred, for which IoU (intersection over union) ≥ threshold.
- False positive (FP): This is a wrong positive detection, that is, a detection for which IoU < threshold.
- False Negative (FN): This is a real incident that the classifier failed to recognize.
- True Negative (TN): Given that the real instance is also negative, this metric suggests a negative detection. TN comprises all possible wrong detection that was not detected.

$$Overall\ error\ rate = \frac{FN + FP}{TP + TN + FP + FN}$$

Displays the confusion matrix for the YOLOv8 model, which consists of two classes: true class and predicted class. The true positives, indicating correct Object predictions, Account for 83 per cent accuracy in the model's performance. Conversely, there is a 17 per cent rate of false negatives, signifying instances where the model failed to recognize object as Object. The true negatives stand at 100 per cent, meaning the model did not mistakenly classify any other objects as Object, achieving perfect accuracy in this aspect.

Object Detection in Image.

References

(2023). Real-time Object Detection Using Deep Learning. Journal of

{Image recognition;TV;Law enforcement;Face

10.9734/jamcs/2023/v38i81787.

2015. 5537-5544. 10.1109/ICRA.2015.7139973.

2022, pp. 1175-1179, doi: 10.1109/ICAC3N56670.2022.10074142. keywords:

A flow-based visual programming language for mobile manipulation tasks.

A. A. Sambhe and A. V. Deorankar, and quot;Face Detection And Recognization

Advances in Mathematics and Computer Science. 38. 24-32.

Alexandrova, Sonya and amp; Tatlock, Zachary and amp; Cakmak, Maya. (2015). RoboFlow: Communication Control and Networking (ICAC3N), Greater Noida, India,

https: //roboflow.com/model/yolov8

Learning. 10.1007/s40009-023-01265-4.

Patro, K Suresh and amp; Bharti, Vidya and amp; Sharma, Arun and amp; Sharma, Arpita and amp; Yadav, Proceedings - IEEE International Conference on Robotics and Automation. recognition;Lighting;Cameras;Face detection;Face detection;face recognization;face extraction;pca;knn;haarcascade},

Sohan, Mupparaju and amp; Ram, Thotakura and amp; Ch, Venkata. (2024). A Review on System, and quot; 2022 4th International Conference on Advances in Computing,

Vaishnavi, K. and amp; Reddy, G. and amp; Reddy, T. and amp; Iyengar, N. and amp; Shaik, Subhani. Vinod. (2023). Fish Detection in Underwater Environments Using Deep

YOLOv8 and Its Advancements. 10.1007/978-981-99-7962-2_39.

Chapter 15

Practical on Object Detection using YOLOv8

Program for Adult-Child Detection Using Roboflow-Yolov8 in Google Colab

Dataset

https: //universe.roboflow.com/hua-jdg0z/kids_adult/dataset/5#

Program

train_yolov8_object_detection_on_TRAINING.ipynb

Output

Set Home

```
[1] import os
    HOME = os.getcwd()
    print(HOME)

/content
```

Install Libraries

```
[2]  # Pip install method (recommended)

     !pip install ultralytics==8.0.196

     from IPython import display
     display.clear_output()

     import ultralytics
     ultralytics.checks()

Ultralytics YOLOv8.0.196 🚀 Python-3.10.12 torch-2.3.0+cu121 CUDA:0 (Tesla T4, 15102MiB)
Setup complete ✅ (2 CPUs, 12.7 GB RAM, 30.2/78.2 GB disk)
```

Import Libraries

```
[3]  from ultralytics import YOLO

     from IPython.display import display, Image
```

Make Dataset folder Then install roboflow and get roboflow dataset in this folder:

```
!mkdir {HOME}/datasets
%cd {HOME}/datasets

!pip install roboflow

from roboflow import Roboflow
rf = Roboflow(api_key="Gw13eiUUHdygQL32vRe3")
project = rf.workspace("worktest-qsylv").project("child-adult-detection")
version = project.version(2)
dataset = version.download("yolov8")

/content/datasets
Collecting roboflow
  Downloading roboflow-1.1.33-py3-none-any.whl (75 kB)
     ━━━━━━━━━━━━━━━━━━━━━━━━━━━━━━━━━━━━━━━━ 75.6/75.6 kB 3.4 MB/s eta 0:00:00
Requirement already satisfied: certifi in /usr/local/lib/python3.10/dist-packages (from roboflow) (2024.6.2)
Collecting chardet==4.0.0 (from roboflow)
  Downloading chardet-4.0.0-py2.py3-none-any.whl (178 kB)
     ━━━━━━━━━━━━━━━━━━━━━━━━━━━━━━━━━━━━━━━━ 178.7/178.7 kB 13.2 MB/s eta 0:00:00
Requirement already satisfied: idna==3.7 in /usr/local/lib/python3.10/dist-packages (from roboflow) (3.7)
Requirement already satisfied: cycler in /usr/local/lib/python3.10/dist-packages (from roboflow) (0.12.1)
Requirement already satisfied: kiwisolver>=1.3.1 in /usr/local/lib/python3.10/dist-packages (from roboflow) (1.4.5)
Requirement already satisfied: matplotlib in /usr/local/lib/python3.10/dist-packages (from roboflow) (3.7.1)
Requirement already satisfied: numpy>=1.18.5 in /usr/local/lib/python3.10/dist-packages (from roboflow) (1.25.2)
Requirement already satisfied: opencv-python-headless==4.10.0.84 in /usr/local/lib/python3.10/dist-packages (from roboflow) (4.10.0.84)
Requirement already satisfied: Pillow>=7.1.2 in /usr/local/lib/python3.10/dist-packages (from roboflow) (9.4.0)
Requirement already satisfied: python-dateutil in /usr/local/lib/python3.10/dist-packages (from roboflow) (2.8.2)
Collecting python-dotenv (from roboflow)
  Downloading python_dotenv-1.0.1-py3-none-any.whl (19 kB)
Requirement already satisfied: requests in /usr/local/lib/python3.10/dist-packages (from roboflow) (2.31.0)
Requirement already satisfied: six in /usr/local/lib/python3.10/dist-packages (from roboflow) (1.16.0)
Requirement already satisfied: urllib3>=1.26.6 in /usr/local/lib/python3.10/dist-packages (from roboflow) (2.0.7)
Requirement already satisfied: tqdm>=4.41.0 in /usr/local/lib/python3.10/dist-packages (from roboflow) (4.66.4)
Requirement already satisfied: PyYAML>=5.3.1 in /usr/local/lib/python3.10/dist-packages (from roboflow) (6.0.1)
Collecting requests-toolbelt (from roboflow)
  Downloading requests_toolbelt-1.0.0-py2.py3-none-any.whl (54 kB)
     ━━━━━━━━━━━━━━━━━━━━━━━━━━━━━━━━━━━━━━━━ 54.5/54.5 kB 3.2 MB/s eta 0:00:00
Collecting python-magic (from roboflow)
  Downloading python_magic-0.4.27-py2.py3-none-any.whl (13 kB)
Requirement already satisfied: contourpy>=1.0.1 in /usr/local/lib/python3.10/dist-packages (from matplotlib->roboflow) (1.2.1)
Requirement already satisfied: fonttools>=4.22.0 in /usr/local/lib/python3.10/dist-packages (from matplotlib->roboflow) (4.53.0)
Requirement already satisfied: packaging>=20.0 in /usr/local/lib/python3.10/dist-packages (from matplotlib->roboflow) (24.1)
Requirement already satisfied: pyparsing>=2.3.1 in /usr/local/lib/python3.10/dist-packages (from matplotlib->roboflow) (3.1.2)
Requirement already satisfied: charset-normalizer<4,>=2 in /usr/local/lib/python3.10/dist-packages (from requests->roboflow) (3.3.2)
Installing collected packages: python-magic, python-dotenv, chardet, requests-toolbelt, roboflow
  Attempting uninstall: chardet
    Found existing installation: chardet 5.2.0
    Uninstalling chardet-5.2.0:
      Successfully uninstalled chardet-5.2.0
Successfully installed chardet-4.0.0 python-dotenv-1.0.1 python-magic-0.4.27 requests-toolbelt-1.0.0 roboflow-1.1.33
WARNING: The following packages were previously imported in this runtime:
  [chardet]
You must restart the runtime in order to use newly installed versions.
RESTART SESSION
loading Roboflow workspace...
loading Roboflow project...
Downloading Dataset Version Zip in Child-Adult-Detection-2 to yolov8:: 100%|██████████| 545002/545002 [00:32<00:00, 16723.84it/s]

Extracting Dataset Version Zip to Child-Adult-Detection-2 in yolov8:: 100%|██████████| 13596/13596 [00:03<00:00, 4460.03it/s]
```

Train the dataset as its epoach times and get the train-weights.

```
[5] %cd {HOME}

    !yolo task=detect mode=train model=yolov8s.pt data={dataset.location}/data.yaml epochs=50 imgsz=640 plots=True

              Class     Images  Instances      Box(P          R      mAP50  mAP50-95): 100% 21/21 [00:08<00:00,  2.35it/s]
                all        644       1261      0.758       0.72      0.745      0.527

      Epoch    GPU_mem   box_loss   cls_loss   dfl_loss  Instances       Size
      43/50      4.13G     0.5802     0.4032      1.081         21        640: 100% 360/360 [02:26<00:00,  2.45it/s]
              Class     Images  Instances      Box(P          R      mAP50  mAP50-95): 100% 21/21 [00:09<00:00,  2.19it/s]
                all        644       1261      0.757      0.721      0.739      0.533

      Epoch    GPU_mem   box_loss   cls_loss   dfl_loss  Instances       Size
      44/50      4.13G     0.5707     0.3931      1.073         22        640: 100% 360/360 [02:28<00:00,  2.42it/s]
              Class     Images  Instances      Box(P          R      mAP50  mAP50-95): 100% 21/21 [00:07<00:00,  2.72it/s]
                all        644       1261      0.742      0.744      0.747      0.539

      Epoch    GPU_mem   box_loss   cls_loss   dfl_loss  Instances       Size
      45/50      4.13G     0.5587     0.3775      1.063         15        640: 100% 360/360 [02:23<00:00,  2.50it/s]
              Class     Images  Instances      Box(P          R      mAP50  mAP50-95): 100% 21/21 [00:09<00:00,  2.12it/s]
                all        644       1261      0.784      0.703      0.752      0.544

      Epoch    GPU_mem   box_loss   cls_loss   dfl_loss  Instances       Size
      46/50      4.15G       0.55     0.3692      1.055         23        640: 100% 360/360 [02:19<00:00,  2.58it/s]
              Class     Images  Instances      Box(P          R      mAP50  mAP50-95): 100% 21/21 [00:10<00:00,  1.95it/s]
                all        644       1261      0.748      0.728      0.748      0.543

      Epoch    GPU_mem   box_loss   cls_loss   dfl_loss  Instances       Size
      47/50      4.13G     0.5426     0.3613      1.052         17        640: 100% 360/360 [02:21<00:00,  2.55it/s]
              Class     Images  Instances      Box(P          R      mAP50  mAP50-95): 100% 21/21 [00:09<00:00,  2.28it/s]
                all        644       1261      0.803      0.698      0.756       0.55

      Epoch    GPU_mem   box_loss   cls_loss   dfl_loss  Instances       Size
      48/50      4.13G     0.5271      0.349      1.042         17        640: 100% 360/360 [02:24<00:00,  2.49it/s]
              Class     Images  Instances      Box(P          R      mAP50  mAP50-95): 100% 21/21 [00:07<00:00,  2.65it/s]
                all        644       1261      0.776      0.723      0.756      0.553

      Epoch    GPU_mem   box_loss   cls_loss   dfl_loss  Instances       Size
      49/50      4.13G     0.5165     0.3381      1.034         18        640: 100% 360/360 [02:19<00:00,  2.57it/s]
              Class     Images  Instances      Box(P          R      mAP50  mAP50-95): 100% 21/21 [00:07<00:00,  2.78it/s]
                all        644       1261      0.766       0.71      0.739      0.541

      Epoch    GPU_mem   box_loss   cls_loss   dfl_loss  Instances       Size
      50/50      4.13G     0.5094     0.3326      1.028         19        640: 100% 360/360 [02:25<00:00,  2.47it/s]
              Class     Images  Instances      Box(P          R      mAP50  mAP50-95): 100% 21/21 [00:07<00:00,  2.74it/s]
                all        644       1261      0.808      0.694      0.754       0.55

50 epochs completed in 2.230 hours.
Optimizer stripped from runs/detect/train/weights/last.pt, 22.5MB
Optimizer stripped from runs/detect/train/weights/best.pt, 22.5MB

Validating runs/detect/train/weights/best.pt...
Ultralytics YOLOv8.0.196 🚀 Python-3.10.12 torch-2.3.0+cu121 CUDA:0 (Tesla T4, 15102MiB)
Model summary (fused): 168 layers, 11126358 parameters, 0 gradients, 28.4 GFLOPs
              Class     Images  Instances      Box(P          R      mAP50  mAP50-95): 100% 21/21 [00:14<00:00,  1.48it/s]
                all        644       1261      0.775      0.724      0.756      0.553
              Adult        644        372      0.629      0.573      0.576      0.304
              Child        644        889      0.921      0.875      0.935      0.802
Speed: 0.3ms preprocess, 4.9ms inference, 0.0ms loss, 3.7ms postprocess per image
Results saved to runs/detect/train
💡 Learn more at https://docs.ultralytics.com/modes/train
```

After training get dataset confusion matrix, its accuracy.

```
[6] !ls {HOME}/runs/detect/train/

args.yaml                                          R_curve.png               val_batch0_labels.jpg
confusion_matrix_normalized.png                    results.csv               val_batch0_pred.jpg
confusion_matrix.png                               results.png               val_batch1_labels.jpg
events.out.tfevents.1719893243.0cf0261ee865.3318.0 train_batch0.jpg          val_batch1_pred.jpg
F1_curve.png                                       train_batch14400.jpg      val_batch2_labels.jpg
labels_correlogram.jpg                             train_batch14401.jpg      val_batch2_pred.jpg
labels.jpg                                         train_batch14402.jpg      weights
P_curve.png                                        train_batch1.jpg
PR_curve.png                                       train_batch2.jpg
```

Displaying the Confusion matrix

```
%cd {HOME}
Image(filename=f'{HOME}/runs/detect/train/confusion_matrix.png', width=600)
```

/content

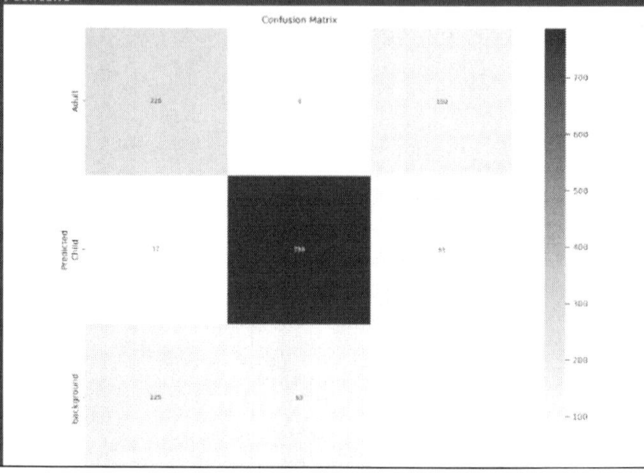

Displaying the curve Results of the loss and mAP values.

```
[8] %cd {HOME}
Image(filename=f'{HOME}/runs/detect/train/results.png', width=600)
```

/content

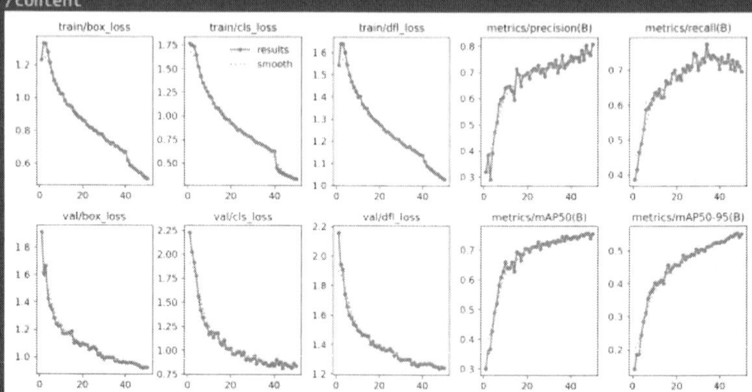

Displaying the Validation images with its accuracies.

```
[9] %cd {HOME}
    Image(filename=f'{HOME}/runs/detect/train/val_batch0_pred.jpg', width=600)
```

```
/content
```

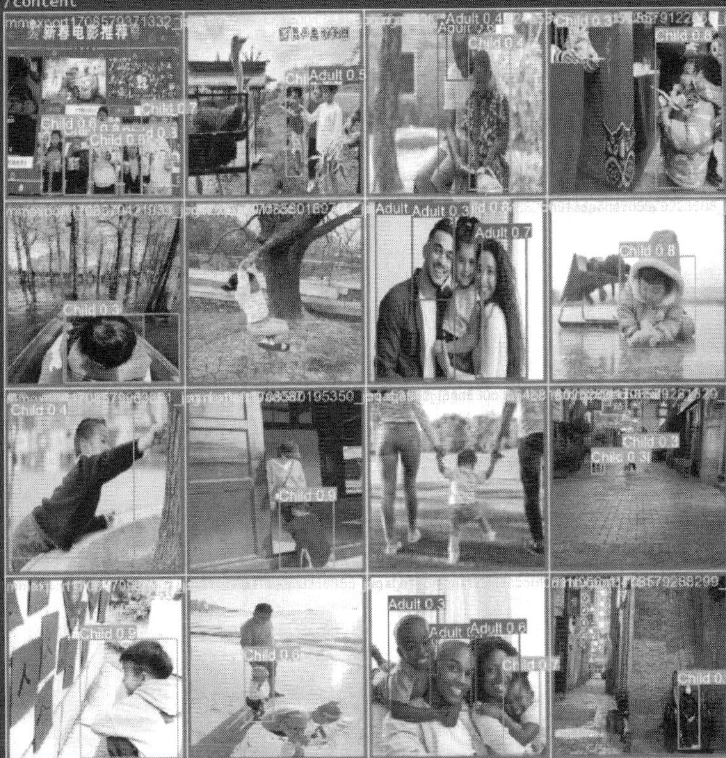

Displaying the Validation result with its accuracies.

```
[10] %cd {HOME}

     !yolo task=detect mode=val model={HOME}/runs/detect/train/weights/best.pt data={dataset.location}/data.yaml
```

```
/content
Ultralytics YOLOv8.0.196 🚀 Python-3.10.12 torch-2.3.0+cu121 CUDA:0 (Tesla T4, 15102MiB)
Model summary (fused): 168 layers, 11126358 parameters, 0 gradients, 28.4 GFLOPs
val: Scanning /content/datasets/Child-Adult-Detection-3/valid/labels.cache... 644 images, 0 backgrounds, 0 corrupt: 100% 644/644 [00:00<?, ?it/s]
WARNING ⚠ Box and segment counts should be equal, but got len(segments) = 2, len(boxes) = 1261. To resolve this only boxes will be used and all segments will be removed. To avoid this
                 Class     Images  Instances      Box(P          R      mAP50  mAP50-95): 100% 41/41 [00:13<00:00,  3.06it/s]
                   all        644       1261      0.776      0.724      0.756      0.553
                 Adult        644        372       0.63      0.571      0.577      0.304
                 Child        644        889      0.922      0.875      0.935      0.802
Speed: 1.3ms preprocess, 9.4ms inference, 0.0ms loss, 2.7ms postprocess per image
Results saved to runs/detect/val
💡 Learn more at https://docs.ultralytics.com/modes/val
```

Give the Unknown image path to trained weight for detecting the objects in the given images.

```
[16] %cd {HOME}
     !yolo task=detect mode=predict model={HOME}/runs/detect/train/weights/best.pt conf=0.25 source=/content/ad11.jpg save=True

/content
Ultralytics YOLOv8.0.196 🚀 Python-3.10.12 torch-2.3.0+cu121 CUDA:0 (Tesla T4, 15102MiB)
Model summary (fused): 168 layers, 11126358 parameters, 0 gradients, 28.4 GFLOPs

WARNING ⚠ NMS time limit 0.550s exceeded
image 1/1 /content/ad11.jpg: 288x640 2 Adults, 1 Child, 155.2ms
Speed: 3.4ms preprocess, 155.2ms inference, 815.3ms postprocess per image at shape (1, 3, 288, 640)
Results saved to runs/detect/predict3
💡 Learn more at https://docs.ultralytics.com/modes/predict
```

Displaying the Given input images with its Object Detection output.

```
[17] import glob
     from IPython.display import Image, display

     for image_path in glob.glob(f'{HOME}/runs/detect/predict3/*.jpg')[:3]:
         display(Image(filename=image_path, width=600))
         print("\n")
```

Index

E

F

G

H

I

K

L

U

V

W

Y